Springer Tracts in Modern Physics 82

Ergebnisse der exakten Naturwissenschaften

Springer Tracts in Modern Physics

Volume 66	Quantum Statistics in Optics and Solid-State Physics
Volume 67	Conformal Algebra in Space-Time and Operator Product Expansion
Volume 68	Solid-State Physics
Volume 69	Astrophysics
Volume 70	Quantum Optics
Volume 71	Nuclear Physics
Volume 72	Van der Waals Attraction
Volume 73	Excitons at High Density
Volume 74	Solid-State Physics
Volume 75	Light Scattering by Phonon-Polaritons
Volume 76	Irreversible Properties of Type II Superconductors
Volume 77	Surface Physics
Volume 78	Solid-State Physics
Volume 79	Elementary Particle Physics
Volume 80	Neutron Physics
Volume 81	Point Defects in Metals I: Introduction to the Theory
Volume 82	Electronic Structure of Noble Metals, and Polariton-Mediated Light Scattering

Electronic Structure of Noble Metals
and
Polariton-Mediated Light Scattering

Contributions by
B. Bendow B. Lengeler

With 42 Figures

Springer-Verlag
Berlin Heidelberg New York 1978

Dr. Bernard Bendow

Rome Air Development Center, Deputy for Electronic Technology,
Hanscom AFB, MA 01731, USA

Dr. Bruno Lengeler

Institut für Festkörperforschung der Kernforschungsanlage Jülich
Postfach 1913, D-5170 Jülich
(Present address: Bell Laboratories, 600 Mountain Avenue, Murray Hill, NJ 07974, USA)

Manuscripts for publication should be addressed to:

Gerhard Höhler

Institut für Theoretische Kernphysik der Universität Karlsruhe
Postfach 6380, D-7500 Karlsruhe 1

Proofs and all correspondence concerning papers in the process of publication should be addressed to:

Ernst A. Niekisch

Institut für Grenzflächenforschung und Vakuumphysik der Kernforschungsanlage Jülich
Postfach 1913, D-5170 Jülich

ISBN 3-540-08814-8 Springer-Verlag Berlin Heidelberg New York
ISBN 0-387-08814-8 Springer-Verlag New York Heidelberg Berlin

Library of Congress Cataloging in Publication Data. Bendow, Bernard, 1942 —. Electronic structure of noble metals and polariton-mediated light scattering. (Springer tracts in modern physics; v. 82) Bibliography: p. Includes index. 1. Polaritons. 2. Precious metals. 3. Electronic structure. I. Lengeler, B., 1939 —. joint author. II. Title. III. Series. QC1.S797 vol. 82 [QC176.8.P6] 539'.08s [530.4'1] ISBN 0-387-08814-8 78-18848

Offset printing and bookbinding: Brühlsche Universitätsdruckerei, Lahn-Giessen
2153/3130 — 543210

Contents

de Haas-van Alphen Studies of the Electronic Structure of the Noble Metals and Their Dilute Alloys

By *B. Lengeler*. With 26 Figures

Polariton Theory of Resonance Raman Scattering in Solids

By *B. Bendow*. With 16 Figures

de Haas-van Alphen Studies of the Electronic Structure of the Noble Metals and Their Dilute Alloys

Bruno Lengeler

1. Introduction

The de-Haas-van Alphen (dHvA) effect is one of the quantum oscillation phenomena which are characterized by the redistribution of conduction electron states on Landau cylinders in a magnetic field. The Landau cylinders expand with increasing field and leave the Fermi surface one by one. As a result, the electronic density of states at the Fermi level changes periodically with the field. Thus quantum oscillations occur in all physical quantities which contain the density of states. Among these are the magnetoresistance, the Hall effect, the thermoelectric effect, the contact potential between two metals, the electronic specific heat, and the ultrasonic absorption in metals. The dHvA effect is the quantum oscillation of the magnetization of the conduction electrons. The effect has evolved from a curiosity, first observed in bismuth, to one of the most powerful methods for the investigation of the electronic structure of pure metals, intermetallic compounds, and dilute alloys. The geometry of the Fermi surface can be deduced from the frequencies of the dHvA oscillations, and the Fermi surfaces of nearly all pure metals and of many ordered alloys have been determined in this manner. These measurements have had a great influence on our understanding of the electronic structure of metals. In more recent years, the interest has shifted towards the information contained in the dHvA amplitudes. Cyclotron masses and Fermi velocities can be derived from the temperature dependence of these amplitudes, whereas their field dependence determines the Dingle temperatures and the scattering of the conduction electrons at defects.

Two characteristic features of the dHvA effect, and of quantum oscillations in general, should be emphasized. First, only the electronic states at the Fermi level and in its immediate vicinity can be investigated by the dHvA effect because only those states are affected by the depletion of the Landau levels when a cylinder leaves the Fermi surface. Electronic states which are farther away from the Fermi level than $k_B T$ must be investigated by other methods, for instance, by optical spectroscopy. Nonetheless, the properties of the electronic system at the Fermi level can be measured by dHvA effect with great accuracy. The linear dimensions of the Fermi surface for certain metals are known with an accuracy of 1 part in 10^4 or better. The immediate vicinity of the Fermi surface is also accessible to the dHvA effect. By thermal excitation of the states, a range of width $k_B T$ around E_F can be scanned. Thus the temperature dependence of the dHvA amplitudes contains the gradients on the Fermi surface, i.e., the cyclotron masses and the Fermi velocities. A second characteristic feature of the dHvA effect is that all quantities derived from the dHvA effect are averages over the extremal cross section on the Fermi surface for the field direction under consideration. Thus the dHvA frequency is an orbital average of the radii of the extremal cross section. The cyclotron masses are orbital averages of the Fermi velocities and the Dingle temperatures are orbital averages of the electron scattering rates. Since only a relatively small number of states at the Fermi level are involved in the effect at one time, local values of the radii, Fermi velocities, and scattering rates can be obtained by measuring the orientation dependence of the orbital averages and by deconvoluting them. The applicability of this procedure is one of the great advantages of the dHvA effect.

The present paper is concerned with the investigation of the electronic structure of the noble metals - copper, silver, and gold - and their dilute alloys by means of the dHvA effect. The paper is organized as follows. In Section 2, the dHvA effect is explained in a semiclassical way and the Lifshitz-Kosevich expression for the oscillatory magnetization is given. The various applications of the dHvA effect are described, and the influence of the electron-phonon interaction on the dHvA effect is treated at some length. In Section 3, details are given of the field modulation technique by which most of the frequency and amplitude measurements have been made. In Section 4, a detailed description of the geometry of the Fermi surface of the noble metals is given. The anisotropy of the Fermi surface is explained within a band structure calculation by the hybridization of the s-, p-, and d-bands. Section 5 gives a description of detailed cyclotron mass measurements in Cu, Ag, and Au. From these data are derived values of the Fermi velocities and of the coefficient γ^* of the specific heat. For Cu, the anisotropy of the electron-phonon coupling constant is obtained by comparing the Fermi velocities derived from cyclotron masses with those obtained from a band structure calculation. Finally, in Section 6 measurements of Dingle temperatures for some dilute alloys of the noble metals are discussed. Only alloys in which the electron scattering is spin independent are considered. The influence of the scattering strength of the defect, of its position in the lattice, and of the wave character of the conduction electrons on the observed scattering rates is explained in detail by means of a generalized phase shift analysis.

2. The de Haas-van Alphen (dHvA) Effect

In 1930, DE HAAS and VAN ALPHEN observed that the susceptibility of single crystal bismuth varied at low temperature in an oscillatory way with the magnetic field /2.1/. The amplitude of the oscillations decreased with increasing temperature and the effect disappeared at about 35 K. PEIERLS correlated the oscillations with the quantization of the orbits of the free conduction electrons in the magnetic field /2.2/. The first explicit expression for the variation of the magnetization with the field was given by LANDAU for ellipsoidal energy surfaces /2.3/. ONSAGER showed that the frequency of the dHvA oscillations for arbitrary energy surfaces is proportional to the extremal cross section of the Fermi surface for a given field direction /2.4/. LIFSHITZ and KOSEVICH have extended LANDAU's expression for the field dependence of the magnetization for arbitrary energy surfaces /2.5/. Pioneering work in the determination of Fermi surfaces of metals has been done by SHOENBERG /2.6/. To date, the Fermi surfaces of nearly all pure metals and of many ordered compounds have been determined /2.7/. In more recent years, the dHvA effect has been used to determine cyclotron masses, Fermi velocities, and scattering rates of conduction electrons at defects.

Typical dHvA oscillations in gold and copper are shown schematically in Fig. 2.1. When the magnetic field is parallel to a <100> crystallographic direction, the magnetization contains two periodic contributions (Fig. 2.1a). dHvA oscillations can also be observed if the crystal is rotated in a constant magnetic field. Fig. 2.1b shows the oscillations observed if a Cu crystal is turned around an axis <110>.

2.1 Lifshitz-Kosevich Expression for the dHvA Effect

The oscillatory variation of the magnetization of the conduction electrons is described quantitatively be the Lifshitz-Kosevich theory of the dHvA effect. In this theory, the free energy of the conduction electrons is calculated for arbitrary energy surfaces as a function of the magnetic field H. The magnetization of the conduction electrons in a single crystal contains an oscillatory part M which can be determined from the oscillatory part of the free energy G according to

$$\underline{M} = -\partial G/\partial \underline{H} \ . \qquad (2.1)$$

The period of the oscillations is correlated with the extremal cross sections of the Fermi surface. The temperature and field dependence of the amplitudes of the oscillations is determined by the cyclotron masses and by the electron lifetimes. There exist a number of review articles in which the Lifshitz-Kosevich expression of the dHvA effect is presented. An excellent review has been given by GOLD /2.8/. In this paper, we confine ourselves to a representation in which the major physical aspects of the dHvA effect are derived in a semiclassical way.

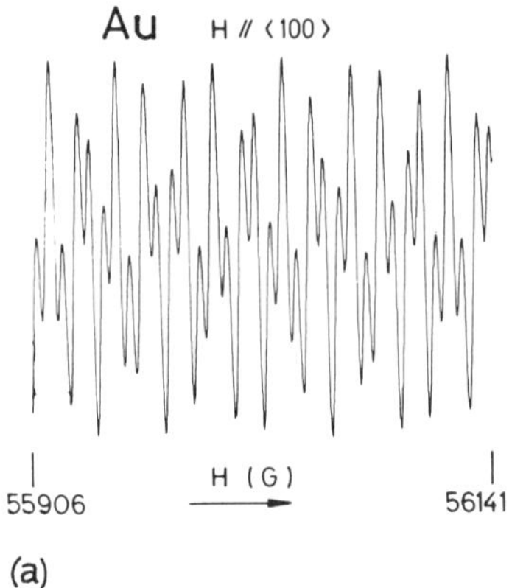

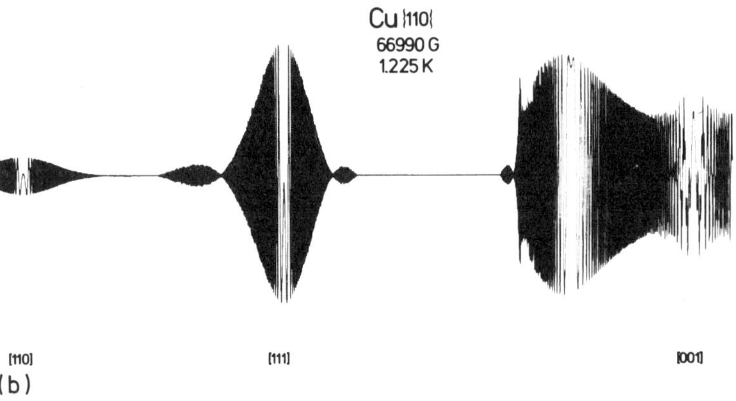

Fig. 2.1a and b. dHvA oscillations in gold. (a) Field dependence of the oscillations at T = 1.179 K. The magnetic field is parallel to a crystallographic direction <100>. The magnetization contains two contributions (B <100> and R <100>). (b) Angular dependence of the oscillations. The crystal is rotated in a constant field through 100^0 around an axis <110>

2.1.1 Conduction Electrons in a Homogeneous Magnetic Field

In the absence of a magnetic field, the conduction electron states in an ideal metallic single crystal are characterized by the wave vectors $\underline{k}$ and the spin quantum numbers s. The $\underline{k}$-vectors specifying the different electronic states are confined to the first Brillouin zone. At $T=0$, all states with energies up to the Fermi energy E_F are occupied. The Fermi surface separates the occupied and unoccupied states. Each state $(\underline{k},s)$ can be occupied only once.

A magnetic field $\underline{H}$ redistributes the possible electronic states and alters their degeneracy. The Lorentz force describes how a state $\underline{k}$ changes in time with the magnetic field

$$\hbar\dot{\underline{k}} = -(e_0/c)\underline{v}\times\underline{B} \cong -(e_0/c)\underline{v}\times\underline{H} \qquad (2.2)$$

where e_0 is the charge of the proton, c is the velocity of light, and $\hbar$ is Planck's constant. $\underline{v}(\underline{k})$ is the velocity of an electron in the state $\underline{k}$. The field experienced by the electrons is the magnetic induction $\underline{B} = \underline{H} + 4\pi(\underline{M}_0 + \underline{M})$, where $\underline{M}_0$ is the nonoscillatory part of the magnetization. Generally, $M_0 + M \ll H$ in nonferromagnetic materials. Therefore B is replaced by H in the following discussion (see also Section 3.5.3). In a stationary magnetic field, an electron moves on a path of constant energy, the cyclotron orbit. Integration of (2.2) gives

$$\underline{k} = -(e_0/\hbar c)\underline{r}\times\underline{H} \ . \qquad (2.3)$$

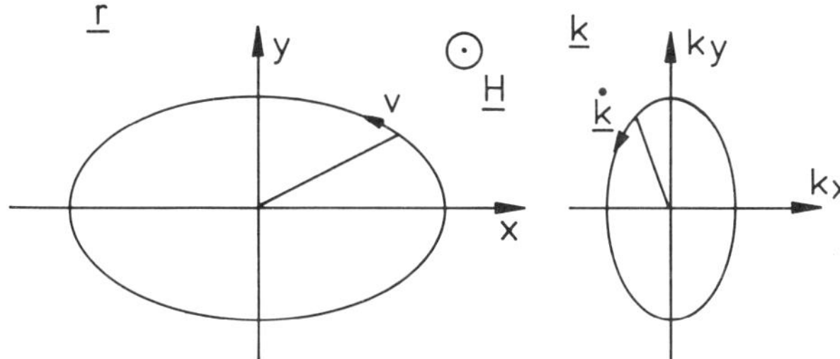

Fig. 2.2. Cyclotron orbits in $\underline{r}$- and in $\underline{k}$-space. Only the projection of the motion on a plane perpendicular to the field is shown

The cyclotron orbit in $\underline{k}$-space is obtained from that in real space by rotation through $\pi/2$ and by scaling with $e_0H/\hbar c$ (Fig.2.2). The electrons move on the cyclotron orbit with the cyclotron frequency ω_c given by

$$\omega_c = e_0H/m_cc \ . \qquad (2.4)$$

The cyclotron mass m_c is obtained in the following manner. According to (2.2) and (2.4) the cyclotron period T_c is

$$T_c = 2\pi/\omega_c = 2\pi c m_c/e_0 H \tag{2.5a}$$

$$= \oint dt = (\hbar c/e_0 H) \oint dk/v_\perp \ . \tag{2.5b}$$

Hence

$$m_c = (\hbar/2\pi) \oint dk/v_\perp \tag{2.6}$$

where $\hbar v_\perp = \delta E/\delta k_\perp$. Therefore

$$m_c = (\hbar^2/2\pi) \oint dk\ \delta k\ /\delta E = (\hbar^2/2\pi)\delta A/\delta E \ . \tag{2.7}$$

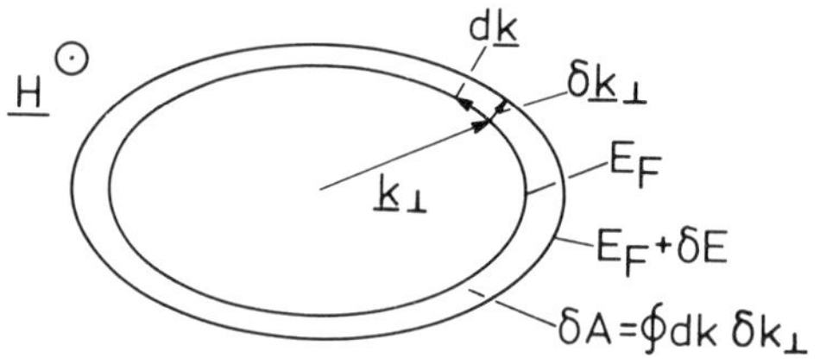

Fig. 2.3. The cyclotron mass is proportional to the energy derivative of the area enclosed by a cyclotron orbit

Fig. 2.3 illustrates the significance of the cyclotron mass which is proportional to the energy derivative of the area enclosed by the cyclotron orbit. For free electrons, m_c reduces to the free electron mass m_0. Using Fig. 2.3, m_c can be written

$$m_c = (\hbar/2\pi) \oint d\alpha\ k^2/(\underline{v}\cdot\underline{k}_\perp) \ . \tag{2.8}$$

The cyclotron orbits are not only lines of constant energy in k-space. In addition, the area enclosed by them must be quantized. The Bohr-Onsager quantization for the projection of the orbit in real space on a plane perpendicular to the magnetic field $\underline{H} = \mathrm{rot}\ \underline{A}$ is written

$$\oint (\hbar\underline{k} - e_0\underline{A}/c)\cdot d\underline{r} = 2\pi\hbar(n+1/2) \ . \tag{2.9}$$

If the areas enclosed by the cyclotron orbit in real space and in $\underline{k}$-space are denoted by S_n and A_n , respectively, (2.9) can be written

$$S_n = \phi_0(n+1/2)/H \tag{2.10}$$

and

$$A_n = 4\pi^2 H(n+1/2)/\phi_0 \ . \tag{2.11}$$

The flux quantum $\phi_o = 2\pi\hbar c/e_o$ has the value $4.1356\cdot10^{-7}$ G cm^2. Eq. (2.11) is of primary importance for the dHvA effect. It indicates that the cyclotron orbits in $\underline{k}$-space expand with the field so that the area enclosed by them increases linearly with H. The energy of the states on the n-th orbit is (disregarding the spin of the electrons and the contributions of the motion parallel to the field),

$$E_n = \hbar\omega_c(n + 1/2) \quad . \tag{2.12}$$

The redistribution of the allowed states by the magnetic field is shown in Fig. 2.4 for free electrons. The quantization parallel to the magnetic field is not altered by the magnetic field. Hence, the electronic states are arranged on a system of cylinders, the Landau cylinders. For free electrons, these are concentric cylinders with circular cross sections and axes parallel to the magnetic field. For arbitrary energy surfaces, the cylinder axis must not coincide with the magnetic field as shown in Fig. 2.5. For any field direction there exists an extremal cross section on the ellipsoid which is perpendicular to the field.

The degeneracy d of the electronic states is also changed by the magnetic field. In the absence of a field, a state $\underline{k}$ can be occupied twice. With a field (H parallel k_z), the degeneracy of the state (n,k_z) is

$$d = L_xL_yH/\phi_o$$

where L_xL_yH is the flux through the crystal cross section L_xL_y perpendicular to the field. d increases linearly with the field H. This is a consequence of the localization of the charge in the field. Whereas a conduction electron in state $\underline{k}$ is spread out through the whole crystal, it is localized on a cylinder with radius v/ω_c in the state (n,k_z). Thus a number of electrons increasing linearly in H can be accommodated in the same state in the same crystal cross section without violating Pauli exclusion.

At T = 0, all those states on the Landau cylinders with energies not exceeding the Fermi energy E_F are occupied. If a large number of Landau cylinders are cut by the Fermi surface, the Fermi energy is somewhat field independent. At increasing field, the effects of decreasing number of Landau cylinders and increasing degeneracy compensate, so that the same number of electrons can be accommodated inside the Fermi surface as at H = 0. If, on the other hand, the field has been increased to such a degree that merely one Landau cylinder is cut by the Fermi surface, E_F must increase with H. This quantum limit is not realized with conventional magnets in the noble metals and will thus be not considered here. In the following, the Fermi energy and the Fermi surface of these metals are treated as field independent.

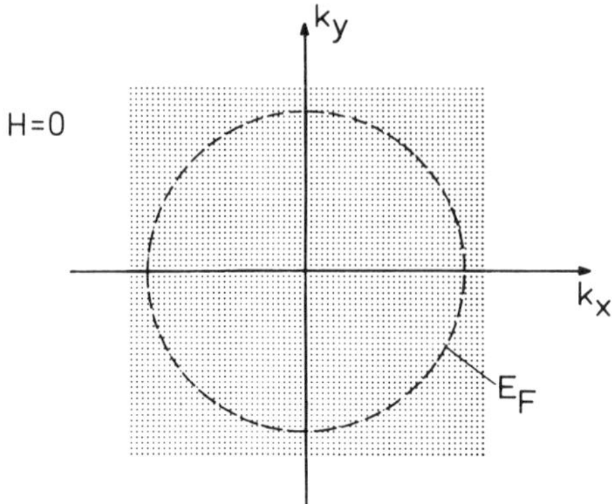

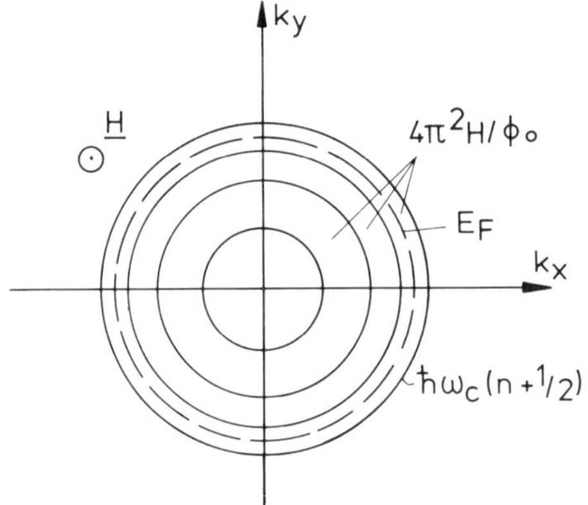

Fig. 2.4. Redistribution of the electronic states in a magnetic field, drawn for free electrons. With magnetic field the states form circles in the planes perpendicular to $\underline{H}$. The area enclosed by neighboring circles increases linearly with the field. The degeneracy of the states increases also linearly with the field. Thus the Fermi energy and the Fermi surface are practically field independent (except for the quantum limit)

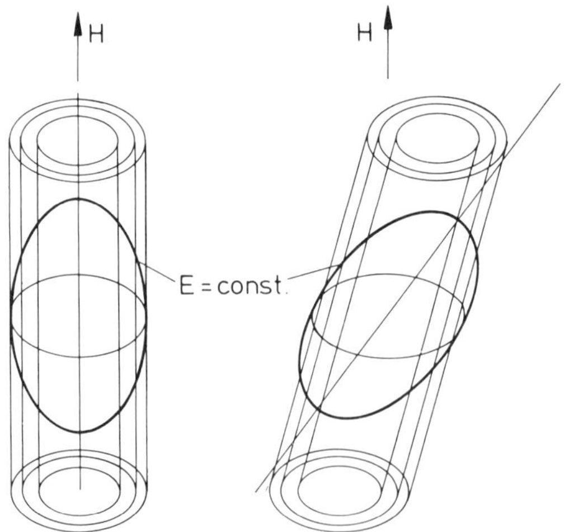

Fig. 2.5. Landau cylinders for ellipsoidal energy surfaces. The cylinder axis coincides with the field direction when a principal axis of the ellipsoid coincides with the field direction

2.1.2 Density of States of the Electrons in the Magnetic Field

The redistribution of the electronic states in $\underline{k}$-space affects the density of states D(E) in a drastic manner. It is obvious from Fig. 2.5 that D(E) has singularities each time a Landau cylinder is tangent to the energy shell. Fig. 2.6 shows the energy spectrum of free electrons with and without field. The density of those states

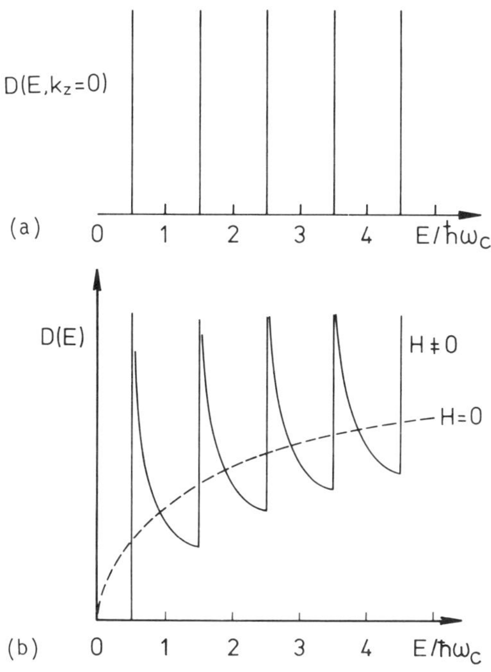

Fig. 2.6a and b. Density of states of free conduction electrons without and with field (b). The density of the states in the extremal cross section is a sum of delta functions separated by $\hbar\omega_c$. Scattering of the conduction electrons at defects causes a Lorentzian broadening of the levels (a)

which are in the extremal cross section is a periodic function of the energy with period $\hbar\omega_c$. For sharp Landau cylinders, it is a sum of delta functions as indicated in Fig. 2.6a. The contribution of the states above and below the extremal cross section to the total density leads to a smooth decrease in D(E) on the upper sides of the singularities. The density of states at the Fermi level changes periodically with increasing field, and drops abruptly each time a Landau cylinder leaves the Fermi surface. This periodic variation of $D(E_F)$ with H is the cause of the quantum oscillations, and in particular, of the dHvA effect.

2.1.3 Origin of the dHvA Oscillations

Figure 2.7 illustrates the origin of the dHvA oscillations. In part (a) it is assumed that at a field strength H_1, the n-th Landau cylinder is tangent to the Fermi

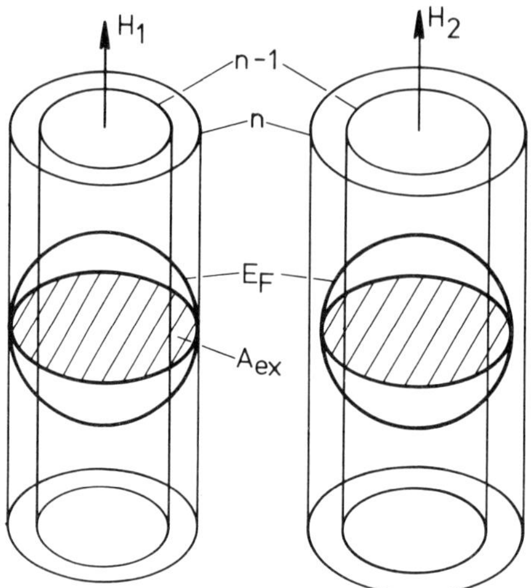

Fig. 2.7. Origin of the dHvA oscillations. Each time a Landau cylinder leaves the Fermi surface with increasing field, the free energy of the electrons drops abruptly. This causes the periodic variations of the magnetization with the field. If the Fermi surface is smeared by finite temperature or if the Landau cylinders are smeared by the scattering of the conduction electrons, the free energy varies less abruptly. Hence the amplitudes of the oscillations are reduced. This argument is valid for arbitrary Fermi surfaces and not only for spherical surfaces as shown here for reasons of simplicity

surface. At T = 0, the Fermi surface is sharp. If the electrons are not scattered at defects, the Landau cylinders are sharp as well. The free energy G of the conduction electrons has a maximum at H_1 because the states on the equator at the Fermi surface are occupied and have the highest energy of all occupied states. If the field is increased from H_1 to H_2, the n-th Landau cylinder leaves the Fermi surface. The states on the equator line of the Fermi surface are depleted and the corresponding electrons are redistributed, mostly on lower energy states. Thus the free energy decreases to a minimum in a small field interval. With further increase of the field, the Landau cylinders further expand, the free energy increases again and reaches another maximum when the (n-1)th cylinder is tangent to the Fermi surface. This completes a cycle of an oscillation of the free energy and of the magnetization of the conduction electrons.

2.1.4 Frequency of the dHvA Oscillations

The frequency of the dHvA oscillations may be deduced immediately from Fig. 2.7. When the n-th Landau cylinder is tangent to the Fermi surface the energy of the states on the contact line is given by

$$\hbar\omega_c(n+1/2) = (\hbar^2/2\pi m_c)A_{ex} \ . \qquad (2.14)$$

Here A_{ex} is the area of the extremal cross section of the Fermi surface for the given field direction. With (2.4), (2.14) can be written

$$F/H = n + 1/2 \qquad (2.15)$$

where

$$F = \phi_0 A_{ex}/4\pi^2 \qquad (2.16)$$

is called the dHvA frequency which is proportional to the extremal cross section A_{ex} at the Fermi surface. The phase of the oscillations is

$$2\pi n = 2\pi(F/H - 1/2) \ . \qquad (2.17)$$

From Fig. 2.7, it is obvious that only the electrons in the extremal cross section of the Fermi surface contribute to the dHvA effect. The density of states in the extremal cross section is a periodic function in energy with period $\hbar\omega_c$ (Fig. 2.6a). It can therefore be expanded in a Fourier series containing terms $\cos(2\pi n)$ and the corresponding higher harmonics. This Fourier series enters the magnetization (2.1) through the free energy G as a sum of harmonics with fundamental $\sin[2\pi(F/H - 1/2)]$. Thus the magnetization varies sinusoidaly with the dHvA period F and its higher harmonics. Since this variation is sinusoidal in 1/H, and not in H, the period ΔH of the oscillations is field dependent

$$\Delta H = H^2/F \ . \qquad (2.18)$$

For a free electron metal with the electron density of gold ($n_{el} = 6\cdot 10^{22}$ cm^{-3}), the values of k_F and F are

$$k_F = (3\pi^2 n_{el})^{1/3} \cong 1.2 \ \text{Å}^{-1} \qquad (2.19)$$

$$F \cong 4.8 \cdot 10^8 \ G \ . \qquad (2.20)$$

At 10^5 G, the dHvA oscillations have a period of $\Delta H \cong 20$ G. For a sample volume of 1 cm^3, there are $2k_F \times 1 \ cm/2\pi \cong 4\times 10^7$ states on a diameter of the Fermi surface at H = 0. In a field of 10^5 G, on the other hand, there are only $n \cong F/H = 4800$ Landau cylinders which are cut by the Fermi surface. The strong diminution of the number of Landau levels and the corresponding increase of the degeneracy of the states in the field imply that a relatively large number of electrons are involved in the depopulation of the Landau cylinders. It is the combination of these two effects which makes the dHvA effect observable experimentally. In the case mentioned above, about $2\cdot 10^{-6}$ of all conduction electrons are involved in the redistribution when a Landau cylinder leaves the Fermi surface.

2.1.5 Amplitude of the dHvA Oscillations

Only those electrons in the extremal cross section of the Fermi surface contribute to the dHvA effect. The thickness of this slice is inversely proportional to the curvature $\hat{\kappa} = \partial^2 A_{ex}/\partial k_H^2$ of the area in the direction of the field. The magnetization of the electrons in the extremal cross-sectional slice is

$$D(H) = -(2e_o\hbar/\pi c m_c)F\phi_o^{-3/2}(H/\hat{\kappa})^{1/2} . \qquad (2.21)$$

It is larger for larger cross sections A_{ex}. D(H) is the maximum value of the amplitude which can only be realized at T = 0 and for sharp Landau cylinders.

Damping of the dHvA Oscillations by Finite Temperature

At finite temperature, the Fermi surface is smeared out according to the Fermi distribution. This smearing implies a less abrupt decrease of the free energy when a Landau cylinder leaves the Fermi surface (Fig. 2.7). Thus, the magnetization $M = -\partial G/\partial H$ is reduced in amplitude. The damping depends on the ratio of the Fermi surface smearing, $k_B T$, and the energy difference, $\hbar\omega_c$, of neighboring Landau cylinders

$$2\pi^2 k_B T/\hbar\omega_c = b m_c T/m_o H . \qquad (2.22)$$

The constant b has the value

$$b = 2\pi^2 k_B c m_o/\hbar e_o = 146.925 \text{ kG/K} . \qquad (2.23)$$

The temperature damping factor I_1 has the form

$$I_1 = (b m_c T/m_o H)[\sinh(b m_c T/m_o H)]^{-1} .$$

At 10^5 G and for $m_c = m_o$, it is $I_1(0\text{ K}) = 1$, $I_1(1\text{ K}) = 0.71$, and $I_1(4.2\text{ K}) = 0.026$. These numbers show that the temperature damping can be appreciable at 4.2 K. This is the reason why the dHvA effect is observable only at low temperatures.

Damping of the dHvA Oscillations by Electron Scattering

Due to the scattering at defects, a conduction electron exists only for a mean time $\tau(\underline{k})$ in a state $\underline{k}$ before it is scattered into another state. According to the uncertainty relation, the scattering broadens the Landau levels. In a phenomenological approach DINGLE has described this broadening by a Lorentzian of width $2\pi k_B X$ /2.9/. X is called Dingle temperature. Since the broadening of the Landau levels reduces the dHvA amplitudes in a way similar to the smearing of the Fermi surface at finite temperature (Fig. 2.7), it is natural to associate it with a temperature, the Dingle temperature. The Dingle damping factor K_1 of the dHvA amplitude depends on the ratio $k_BX/\hbar\omega_c$ which is similar to the ratio (2.22). The density of the Landau levels in the extremal cross section which are broadened according to a Lorentzian (Fig. 2.6a) can again be expanded in a Fourier series. Since the Fourier transform of a Lorentzian is an exponential function, the Dingle damping factor K_1 becomes

$$K_1 = \exp(-bm_cX/m_oH) . \qquad (2.25)$$

MANN has shown how DINGLE'S assumption can be derived from first principles /2.10/. The Dingle temperature is an average of the scattering rates $1/\tau(\underline{k})$ of the electrons over the extremal cross section.

$$X = (\hbar/2\pi k_B) \langle 1/\tau(\underline{k})\rangle . \qquad (2.26)$$

Influence of the Electron Spin on the dHvA Effect

The magnetic moment μ associated with the electron spin $\underline{s}$ can occupy two states in a magnetic field (0,0,H) with the energies

$$\pm g_c\hbar eH/4m_oc = \pm 1/2(g_cm_c/2m_o)\hbar\omega_c . \qquad (2.27)$$

The spin splits the Landau cylinders in two systems of Landau cylinders shifted in energy by the amounts given in (2.27). An electron in a state $(n,k_z = 0,s)$ has the energy

$$E = \hbar\omega_c\ [n + 1/2 \pm 1/2(g_cm_c/2m_o)] . \qquad (2.28)$$

The spin-orbit coupling in a metal can cause deviations of the spin-splitting factor g_c from the value 2 for free electrons. The spin splitting reduces the dHvA amplitude according to the factor

$$S_1 = \cos(\pi g_cm_c/2m_o) . \qquad (2.29)$$

2.1.6 Lifshitz-Kosevich Expression for the dHvA Effect

The main contributions to the magnetization have now been introduced. The oscillatory part of the magnetization parallel to the field is written (neglecting higher dHvA harmonics)

$$M = D(H)I_1K_1S_1\sin[2\pi(F/H - 1/2) \pm \pi/4] . \qquad (2.30)$$

The magnetization varies sinusoidally in 1/H with the dHvA frequency F and the amplitude

$$A(T,H,X) = D(H)I_1K_1S_1 \qquad (2.31)$$

which depends on the temperature, field, Dingle temperature, and orientation of the crystal in the field. It is assumed that the Fermi surface has only one extremal cross section for a given orientation. Otherwise, the contributions from the different cross sections must be added.

2.2 Influence of the Electron-Phonon Interaction on the dHvA Effect

The electrons in a metal can interact with phonons by electron-phonon interactions and with other electrons by Coulomb repulsion and Pauli exclusion. Fig. 2.8 shows how these interactions affect the probability f(E) for an electron to occupy a state

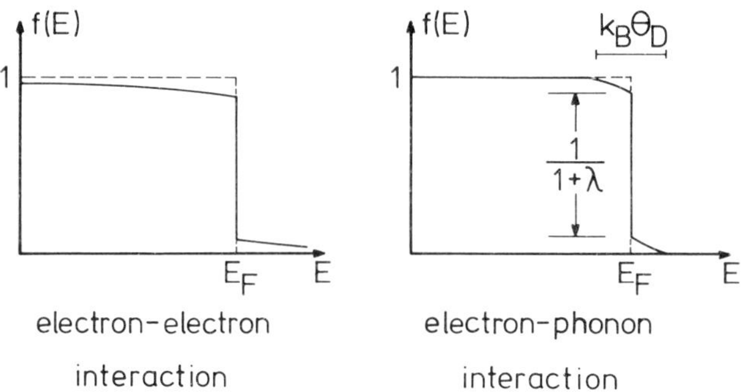

Fig. 2.8. Influence of electron-electron and electron-phonon interaction on the probability f(E) for an electron to occupy a state of energy E

with energy E /2.11/. The interactions smear out the step at E_F in the Fermi distribution (at T = 0). States above E_F are partly occupied and states below E_F are partly depleted. The degree of smearing depends on the strength of the interactions. In the noble metals the average distance of the conduction electrons is $3a_0$ (a_0: Bohr radius). Thus, their mean Coulomb repulsion is $e_0^2/3a_0 \cong 9$ eV. This is more than the Fermi energy. The electron-electron interaction therefore affects even the electrons at the bottom of the conduction band. The electron-phonon interaction, on the other hand, can only affect the electrons in a range of the order $k_B\theta_D$ (θ_D: Debye temperature) around E_F. The electron-phonon interaction reduces the step at E_F from 1 to $1/(1+\lambda)$ where the electron-phonon coupling constant λ for energies near E_F is given by

$$\lambda(\underline{k}) = 2\sum_{\sigma} \int_0^{\omega_{max}} d\omega\ \alpha_\sigma^2(\underline{k},\omega)F_\sigma(\omega)/\omega \ . \tag{2.32}$$

The Eliashberg function $\alpha_{\sigma}^{2}(\underline{k},\omega)F_{\sigma}(\omega)$ describes the coupling of an electron state $\underline{k}$ with all other states by phonons of frequency ω and polarization σ. λ can range from 0.04 to 1.5 depending on the strength of the electron-phonon interaction.

Although the probability f(E) can be affected strongly by the two interactions, the geometry of the Fermi surface is not affected by them. The magnetic field $\underline{H}$ enters the Hamiltonian through the kinetic energy $[\underline{p} + (e_0/c)\underline{A}]$. The quantization of the cyclotron orbits follows from this Hamiltonian, and therefore the period of the dHvA oscillations is independent of the Coulomb and electron-phonon interaction /2.12/. On the other hand, those quantities which contain gradients on the Fermi surface like the Fermi velocities or cyclotron masses are affected by electron-electron and electron-phonon interactions. Realistic band structure calculations take into account the electron-electron interaction whereas they normally do not consider the electron-phonon interaction. A comparison of cyclotron masses measured by dHvA effect with those determined from a band structure calculation, therefore, gives the possibility of estimating the electron-phonon coupling constant $\lambda(\underline{k})$. Since the electron-phonon interaction changes the dispersion relation $E(\underline{k})$ of the conduction electrons in a range of width $k_B\theta_D$ at the Fermi level, the Fermi velocity is reduced by the factor $1/(1+\lambda)$ and the density of states is increased by $(1+\bar{\lambda})$. By $\bar{\lambda}$ is denoted the average of $\lambda(\underline{k})$ over the Fermi surface (Fig. 2.9). It should be emphasized again that the value of the Fermi energy and the geometry of the Fermi surface are unaffected, whereas the gradients (like the Fermi velocity or the density of states of the Fermi level) are. Quantities which are renormalized by electron-phonon interaction will be denoted in the following by an asterisk. The influence of the electron-phonon interaction on the amplitude of the dHvA oscillations is shown in Fig. 2.10. Near E_F it is /2.11/

$$E^{*} = E_F + (E - E_F)/(1+\langle\lambda\rangle) \tag{2.33}$$

where $\langle\lambda\rangle$ is the average of $\lambda(\underline{k})$ around the corresponding cross section

$$\langle\lambda\rangle = \left(\oint dk\ \lambda(\underline{k})/v_{\perp}\right)\cdot\left(\oint dk/v_{\perp}\right)^{-1} . \tag{2.34}$$

Since the slope of E^{*} versus E is $1/(1+\langle\lambda\rangle)$ the distance between neighboring Landau levels, the level broadening and spin splitting are reduced by the factor $1/(1+\langle\lambda\rangle)$ by electron-phonon interaction

$$\hbar\omega_c \rightarrow \hbar\omega_c^{*} = \hbar\omega_c/(1+\langle\lambda\rangle) \tag{2.35}$$

$$X \rightarrow X^{*} = X/(1+\langle\lambda\rangle) \tag{2.36}$$

$$g_c \rightarrow g_c^{*} = g_c/(1+\langle\lambda\rangle) . \tag{2.37}$$

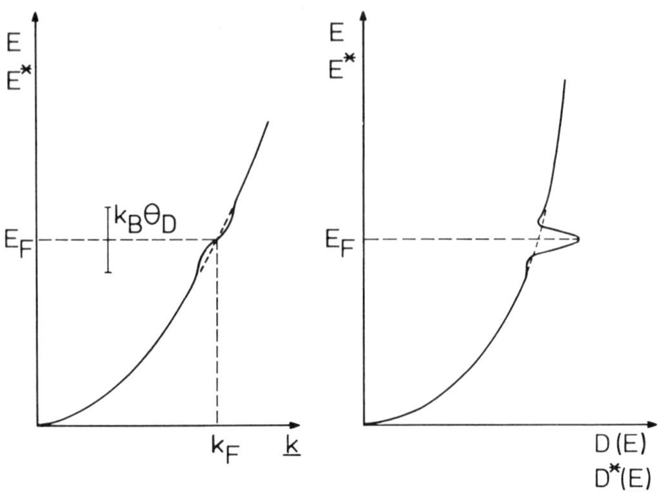

Fig. 2.9. Influence of electron-phonon interaction on the dispersion relation and on the density of states of the electrons

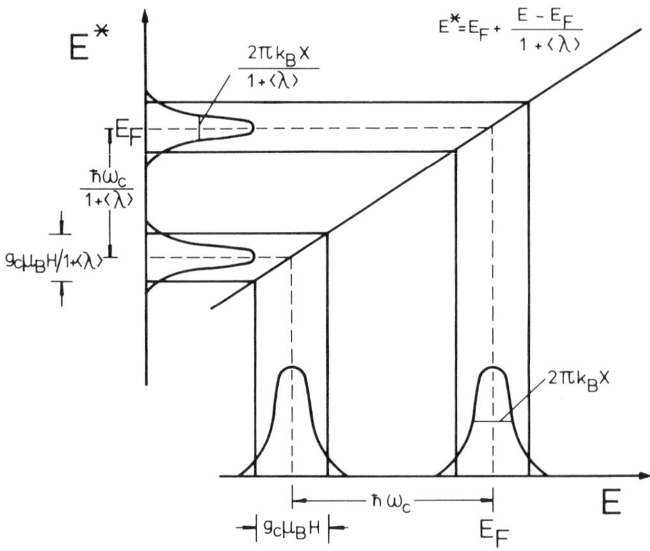

Fig. 2.10. The distance, broadening, and spin splitting of Landau levels are reduced by $1/(1+\langle\lambda\rangle)$ due to electron-phonon interaction. The level broadening has only been drawn for the unsplit levels

The temperature damping of the dHvA amplitudes depends on the ratio of $k_B T$ and the distance between neighboring Landau levels. This ratio now becomes

$$2\pi^2 k_B T/\hbar\omega_c^* = bm_c^* T/m_o H \quad . \tag{2.38}$$

The temperature damping factor I_1 (2.23) therefore contains the cyclotron mass

$$m_c^* = m_c(1 + \langle\lambda\rangle) \tag{2.39}$$

which is enhanced by electron-phonon interaction. In Azbel-Kaner cyclotron resonance, electronic transitions between neighboring Landau levels are induced by an electric rf field. It is therefore again the renormalized mass m_c^* which is determined by this technique.

The Dingle damping factor K_1 (2.25) depends on the ratio $k_B X/\hbar\omega_c$. Since these factors are both reduced by the factor $1/(1 + \langle\lambda\rangle)$ their ratio is independent of the electron-phonon interaction. This can be written in the form

$$m_c X = m_c^* X^* \quad . \tag{2.40}$$

The same argument holds for the spin-splitting factor S_1

$$g_c m_c = g_c^* m_c^* \quad . \tag{2.41}$$

It should be noted that the renormalization of the electronic energies discussed here involves only transitions by virtual phonons. The influence of real phonons on the amplitude of the dHvA effect is not yet completely understood /2.13,14/. In the noble metals, real phonon effects can be neglected at temperatures below 4 K.

2.3 Information Derivable from the dHvA Effect

Including the electron-phonon interaction, the oscillatory part of the magnetization for one extremal cross section and for negligible higher dHvA harmonics is given by

$$M = A(T,H,X^*) \sin[2\pi(F/H - 1/2) \pm \pi/4] \tag{2.42}$$

$$A(T,H,X^*) = D(H) I_1 K_1 S_1 \tag{2.43}$$

$$D(H) = -(2e_o \hbar F/\pi c m_c^* \phi_o{}^{3/2})(H/\hat{\kappa})^{1/2} \tag{2.44}$$

$$I_1 = (bm_c^* T/m_o H)\ [\sinh(bm_c^* T/m_o H)]^{-1} \tag{2.45}$$

$$K_1 = \exp(-bm_c^* X^*/m_o H) \tag{2.46}$$

$$S_1 = \cos(\pi g_c^* m_c^*/2m_o) \tag{2.47}$$

$$F = \phi_o A_{ex}/4\pi^2 . \quad (2.48)$$

All the quantities which can be derived from the dHvA effect are averages of local values around an extremal cross section on the Fermi surface. The thickness of these cross sections depends on the curvature $\hat{\kappa}$ of the Fermi surface. In the noble metals, this thickness is usually of the order of one degree.

2.3.1 Geometry of the Fermi Surface

According to (2.48), the area A_{ex} of an extremal cross section can be derived directly from the corresponding dHvA frequency F. To obtain the linear dimensions of the Fermi surface, it is necessary to measure the angular dependence of the dHvA frequencies. From Fig. 2.3 it can be seen that

$$A_{ex} = 1/2 \oint d\alpha\ k_{\perp}^2 . \quad (2.49)$$

To obtain the geometry of the Fermi surface from (2.49), it is very helpful to have some idea of the shape of the Fermi surface. This can be obtained, for instance, from a band structure calculation. In these calculations, certain parameters (in the electron potential) are fitted in such a way that the calculated cross sections agree optimally with the measured cross sections. In recent years, the Fermi surfaces of most pure metals and of many ordered alloys have been determined by the dHvA effect /2.7/. In some cases, the Fermi surface is known to 1 part in 10^5. This accuracy is achieved at the present time by no other experimental technique. Only for disordered alloys and for those metals for which the preparation of single crystals is difficult is it necessary to use other techniques.

2.3.2 Cyclotron Masses and Fermi Velocities

The cyclotron mass m_c^* for a certain extremal cross section can be obtained from the temperature dependence of the dHvA amplitude A(T) at fixed field H. For suitable experimental conditions, the hyperbolic sine in (2.45) can be replaced by an exponential. A plot of the dHvA amplitude versus the temperature according to

$$\ell n(A/T) = \ell n A_o - (b m_c^*/m_o H)T \quad (2.50)$$

gives a straight line with a slope proportional to m_c^*/m_o. The cyclotron mass is an average of the reciprocal Fermi velocity $v^*(\underline{k})$ over an extremal cross section

$$m_c^* = (\hbar/2\pi) \oint d\alpha\ [k_{\perp}^2/v^*(\underline{k})(\hat{\underline{v}}\cdot\underline{k}_{\perp})]$$

where $\hat{\underline{v}}$ is a unit vector in the direction of the gradient on the Fermi surface. The weighting factor $k_{\perp}^2/(\hat{\underline{v}}\cdot\underline{k}_{\perp})$ depends only on the geometry of the Fermi surface. If this is known, and if the masses have been measured for a sufficient number of cross sections, then a deconvolution of (2.51) gives the local values of the Fermi velocities $v^*(\underline{k})$. The measured values m_c^* and $v^*(\underline{k})$ are renormalized by electron-phonon interaction. If the band velocities $v(\underline{k})$ (not renormalized by electron-phonon interaction) are known from a band structure calculation, local values of the electron--phonon coupling constant $\lambda(\underline{k})$ can be obtained from

$$1 + \lambda(\underline{k}) = v(\underline{k})/v^*(\underline{k}) \; . \tag{2.52}$$

2.3.3 Dingle Temperatures and Scattering Rates of the Conduction Electrons

From measurements of the field dependence of the dHvA amplitudes A(H) at fixed temperatures T, the Dingle temperature X^* can be obtained. A plot

$$\ell n\ [AH^{1/2} \sinh(bm_c^*T/m_oH)] = \ell nA_o - bm_c^*X^*/m_oH \tag{2.53}$$

versus 1/H yields a straight line with slope $-bm_c^*X^*/m_o$. If m_c^* is known for the given extremal cross section, X^* is also known. X^* is the average of the local scattering rates $1/\tau^*(\underline{k})$ of the conduction electrons around the extremal cross section

$$X^* = (\hbar/2\pi k_B)\ \langle 1/\tau^* \rangle \tag{2.54}$$

$$= (\hbar/2\pi)^2(1/k_B m_c^*) \oint dk/(v_\perp^*\tau^*) \tag{2.55}$$

or

$$m_c^*X^* = (\hbar/2\pi)^2(1/k_B) \oint d\alpha\ [k_\perp^2/(\underline{v}^*\cdot\underline{k}_\perp)\tau^*(\underline{k})] \; . \tag{2.56}$$

The weighting factor of the local scattering rate depends on the geometry of the Fermi surface and on the Fermi velocities. The determination of the scattering anisotropy therefore requires a detailed knowledge of the electronic structure of the host lattice in which the scattering defects are distributed. It is not yet completely clear whether the scattering of the electrons by real phonons contributes to the product $m_c^*X^*$ in (2.56) /2.13,14/. However, in the noble metals, only impurities or structural defects contribute to the scattering rates at temperatures below 4.2 K.

2.3.4 g-Factor of the Conduction Electrons

If the absolute values of the dHvA amplitudes are known, the product $g_c^*m_c^*$ can be determined from the spin-splitting factor S_1. This allows the investigation of the $\underline{k}$-dependence of the spin-orbit coupling of the conduction electrons at the Fermi level.

3. Experimental Setup for dHvA Measurements in Cu, Ag, and Au

3.1 Field Modulation Technique

Four experimental techniques have been applied for the measurement of the dHvA effect in the noble metals. These are inductive magnetometers in a pulsed field /3.1/, vibrational magnetometers /3.2/, torque magnetometers /3.3/, and inductive magnetometers with field modulation /3.4/. Since the availability of highly homogeneous and strong superconducting solenoids the field modulation technique has been used in most investigations on noble metals /3.5/. Since this technique has been used in the present investigation the method will be briefly described.

A small alternating field $h \sin \omega t$ of frequency ω is superposed on the large field H_0 which is produced by a superconducting solenoid

$$H(t) = H_0 + h \sin\omega t \ . \tag{3.1}$$

This field creates in a single crystalline metallic sample a magnetization which according to (2.42) is

$$M(t) = A(T,H_0,\dot{X}^*) \sin[2\pi(F/H_0 - 1/2) \pm \pi/4 - 2\pi(Fh/H_0^2)\sin\omega t] \ . \tag{3.2}$$

The high dHvA frequency in the noble metals allows the time dependence of the field in the dHvA amplitude to be neglected relative to that of the phase of the sine. dM/dt induces in a pick-up coil which surrounds the sample a voltage which is proportional to

$$A\sum_{\nu=1}^{\infty} \nu\omega J_\nu(2\pi Fh/H_0^2) \sin(\nu\omega t + \nu\pi/2) \sin[2\pi(F/H_0 - 1/2) \pm \pi/4 + \frac{\nu\pi}{2}] \ . \tag{3.3}$$

Since the magnetization depends in a nonlinear way on the magnetic field, the voltage contains harmonic contributions of the modulation frequency ω which are weighted by Bessel functions J_ν of integer index ν. The harmonic contributions can be filtered out by means of suitable electronic equipment (like, for example, lock-in amplifiers). In most cases, as in the present experiment, the second harmonic is chosen because it gives an optimal signal to noise ratio. The corresponding amplitude is proportional to

$$A(T,H_0,X^*)2\omega J_2(2\pi Fh/H_0^2) \sin[2\pi(F/H_0 - 1/2) \pm \pi/4] \ . \tag{3.4}$$

When the magnetic field H_0 is swept, the dHvA oscillations can be recorded on a XY recorder and Fourier analyzed in subsequent electronic equipment. The amplitude h of the modulation field is chosen in such a way that J_2 is at its first maximum. h must be controlled by the sweep generator to keep $2\pi Fh/H_0^2$ at the value of the first maximum of J_2. To register the dHvA oscillations at a constant rate in time

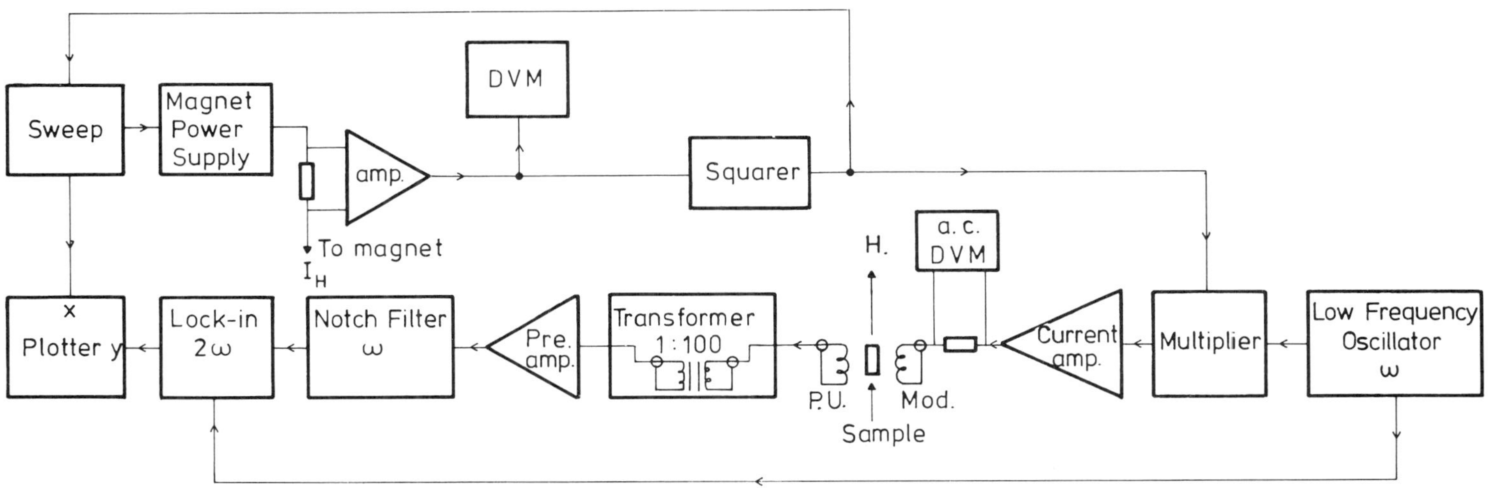

Fig. 3.1. Electronic setup for our dHvA experiment (field modulation technique)

it is necessary to sweep the field at a rate proportional to 1/t. The pick-up coil used was a system of balanced coils one of which contained the sample. A diagram of the electronic setup of the present dHvA experiment is shown in Fig. 3.1.

3.2 Magnet and Cryostat

The superconducting magnet used in this investigation had a maximum field of 76 kG. Since the noble metals have dHvA frequencies of about $5 \cdot 10^8$ G, the period of the oscillations at 50 kG is about 5 G. This imposes strict requirements on both the crystal quality and the homogeneity of the magnetic field. The homogeneity of the magnetic field has been measured by means of an NMR probe (^{27}Al) /3.6/. In the region covered by the sample the field was homogeneous to a few parts in 10^6 over the dimensions of the sample. This is small enough to avoid perceptible reductions of the dHvA amplitudes. The cryostats for the magnet and for the sample holder were conventional bath cryostats. In the inner cryostat which contains the sample, any temperature between 1.1 and 4.2 K can be generated by pumping ^{4}He and kept for hours within a few mK.

3.3 Sample Holder

To measure the angular dependence of the dHvA frequencies, the cyclotron masses and the Dingle temperatures the crystal must be oriented in the field. This is achieved by means of the gear system shown in Fig. 3.2. The sample can be turned around a horizontal axis through 360^0 by means of a worm gear. A second gear system allows tilting by $\pm 5^0$ about an axis perpendicular to the field $\underline{H}$. Part (b) of Fig. 3.2 (which is viewed at 90^0 relative to part (a)) shows the pick-up and the modulation coils. Orientation of a given crystallographic axis along the field is achieved by using the dHvA effect itself. The magnetization is measured as the crystal is turned in the field. Such a measurement is shown in Fig. 2.1. A copper crystal was turned by 100^0 around an axis <110>. When the symmetry directions [001], [111], and [110] are parallel to the field, the magnetization goes through a stationary value. These marks allow any crystallographic direction to be aligned parallel to the field to within one minute of arc. Fig. 3.3 is a detailed view of the angular dependence of the magnetization near <100>. Notice the jerk-free rotation of the crystal in the field at liquid helium temperature.

3.4 Single Crystals of the Noble Metals

The high phase of the dHvA oscillations in the noble metals and the strong angular dependence of the dHvA frequencies necessitate high quality single crystals. Indeed, if a crystal is composed of two grains which are tilted against one another by only

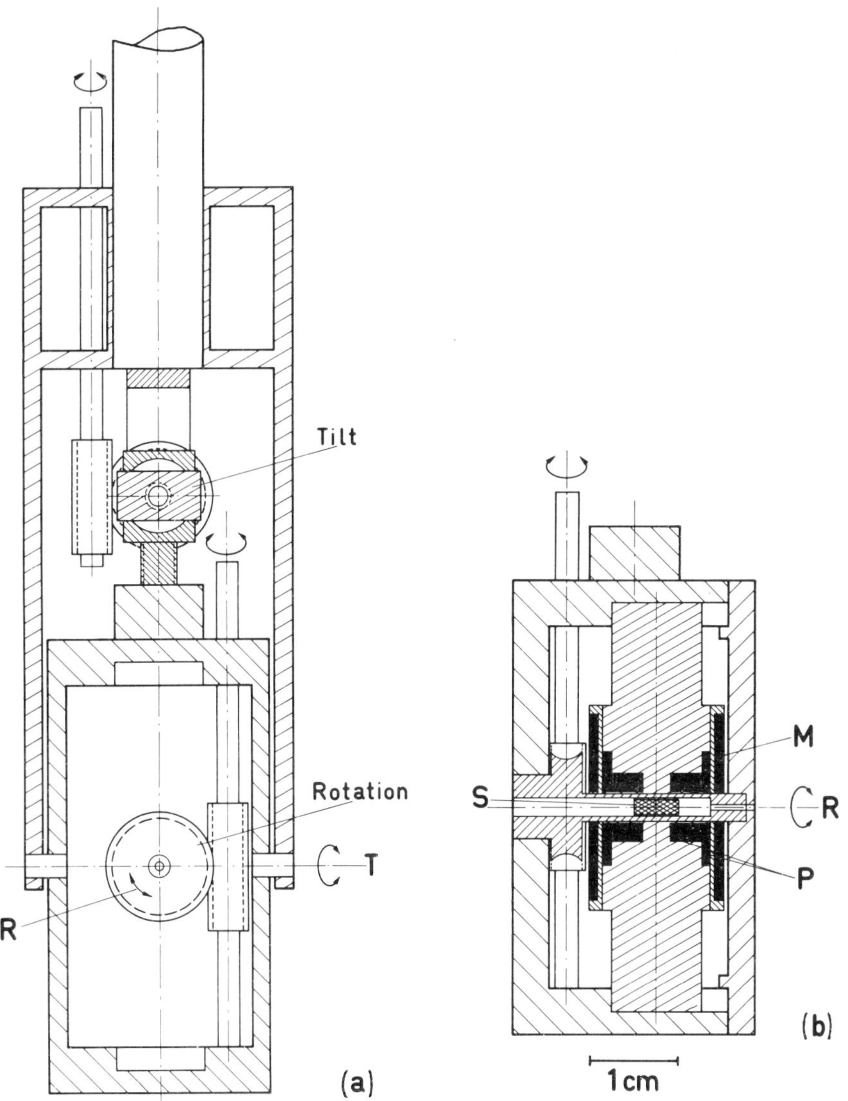

Fig. 3.2a and b. Sample holder used to orient the crystals in the magnetic field. The sample (S) can be rotated through 360^0 around the axis R and tilted by ±5% around the axis T. P and M stand for pick-up and modulation coils

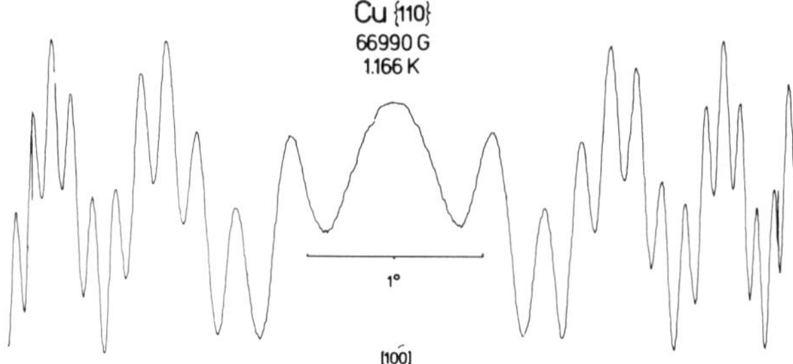

Fig. 3.3. Angular dependence of the magnetization near [100]. The orientation of the crystal in the field can be done within a few minutes of arc

a few minutes of arc, then the dHvA amplitude is markedly reduced for a nonstationary cross section. For that reason the most accurate amplitude measurements can be made for stationary orbits like the high symmetry orbits. High quality single crystals without subgrain boundaries have been grown by the Czochralski technique by UELHOFF and co-workers /3.7/ and were grown as wires 1 mm in diameter and were cut in pieces 5 mm long by an acid layer saw. The width of the rocking curves of these crystals was only 10 - 20 seconds of arc. The dislocation density was as low as 10^3 cm^{-2}. The pure Cu, Ag, and Au crystals had residual resistivities of 0.5, 0.5, and 0.4 nohmcm. Their Dingle temperatures were typically 0.03 K for all orbits.

3.5 Pitfalls in dHvA Measurements

3.5.1 Skin Effect

Accurate amplitude measurements are possible only if the modulation field penetrates homogeneously into the samples. For reasons of signal to noise ratio it was not possible to use modulation frequencies appreciably below 30 Hz in the experimental setup used in this investigation. Although for samples with a resistivity of 0.5 nohmcm the skin depth at 30 Hz is only 0.2 mm, the magnetoresistance in the noble metals increases this skin depth to such a degree that a homogeneous penetration is guaranteed for all orbits except for the belly <100> and the dogsbone <110> orbits. Thus in the pure noble metals, orbits 0.5^0 away from these high symmetry directions have been investigated. Here the magnetoresistance is already large enough. In most doped samples the skin effect creates no problems.

3.5.2 Harmonic dHvA Components

The Lifshitz-Kosevich expression of the dHvA effect contains higher harmonics in the dHvA frequency F in addition to the fundamental term given in (2.42). These components are separated from the fundamental in a Fourier analysis of the dHvA signal. Because they are more strongly damped than the fundamental they can be neglected under suitable experimental conditions. To keep the second harmonic smaller than 1% of the fundamental, the condition $m_c^*(T+X)/m_0 > 1.4$ must be fulfilled at 50 kG. This can be easily achieved.

3.5.3 Magnetic Interaction

It was first emphasized by SHOENBERG /3.1/ that the effective field acting on the electrons is not $\underline{H}$ but $\underline{H}+4\pi(1-D)\underline{M}$. Thus the magnetization (2.42) is given by the implicit equation

$$M = A \sin[2\pi F/|\underline{H}+4\pi(1-D)\underline{M}| \pm \pi/4 - \pi] . \qquad (3.5)$$

D is the demagnetization factor which depends on the geometry of the sample and which is a scalar for ellipsoidally shaped samples. When $4\pi(1-D)M$ is comparable to the dHvA period H^2/F, then appreciable self-modulation occurs which can distort the signal, in an extreme case, from a sine to a sawtooth-shaped curve (magnetic interaction) /3.8/. Deviations of the signal from a pure sine can be most easily monitored by differentiating the signal. Because the magnetic interaction depends on the ratio of the magnetization to the period of the oscillations, it is better to reduce the magnetization by increasing the temperature rather than by decreasing the field. The latter would reduce at the same time the period. The effect is most critical for the belly orbits in the noble metals because they have the highest frequencies and thus the smallest periods.

3.5.4 Phase Smearing

The influence of field and crystal inhomogeneities on the dHvA amplitudes has been pointed out in the Sections 3.2 and 3.4.

4. The Fermi Surface of the Noble Metals

In Section 2.3.1 it was shown how the geometry of the Fermi surface can be deduced from the angular dependence of the dHvA frequencies. The first dHvA measurements of a noble metal were made on Cu by SHOENBERG /4.1/. They showed that the Fermi surface has protrusions in the directions <111> which contact the Brillouin zone. These protrusions form the necks of the Fermi surface of copper. These results confirmed the model of the Fermi surface of Cu proposed by PIPPARD on the basis of anomalous skin effect measurements /4.2/. A graph of the model is shown in Fig. 4.1. In the periodic zone scheme, the Fermi surfaces of the different Brillouin zones are connected at the necks. This leads to both electron orbits (belly B and

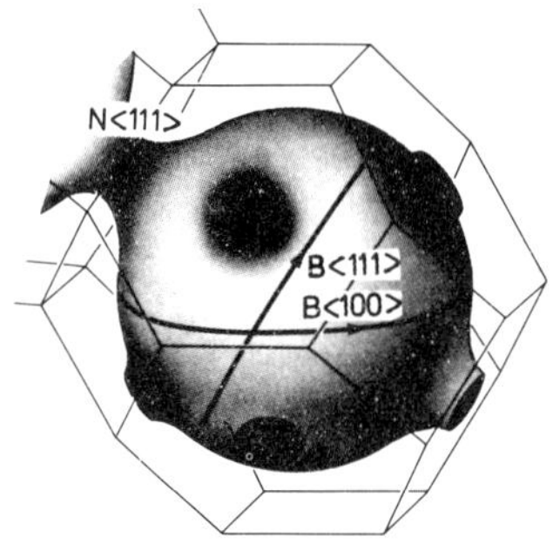

Fig. 4.1. Model of the Fermi surface of the noble metals. The Fermi surface contacts the Brillouin zone in the directions <111>. This gives rise to the necks. Three cyclotron orbits, are shown (the neck orbit N<111> and the two belly orbits B<100> and B<111>)

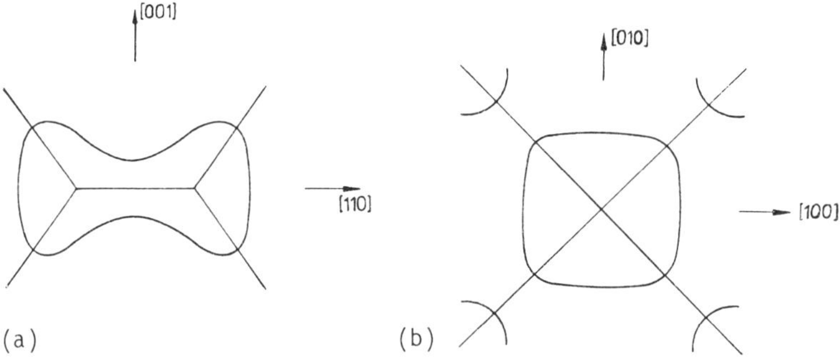

Fig. 4.2a and b. The dogsbone D<110> and the rosette R<100> orbits are observable when the field is parallel to the directions <110> and <100>, respectively

neck N orbits) and to hole orbits with trajectories going through different Brillouin zones. In Fig. 4.2 are shown the dogsbone orbit D<110> when the field is parallel to <110> and the four cornered rosette R<100>. When the magnetic field is parallel to <100>, the Fermi surface has two extremal cross sections, the B<100> and the R<100>. The superposition of the corresponding oscillations in Au is shown in Fig. 2.1. Meanwhile the Fermi surface of the noble metals is among the best investigated of all metals. A detailed reference list up to 1971 is given in the book by CRACKNELL /2.7/. Later very precise measurements on gold were published by BOSACCHI et al. /4.4/. COLERIDGE and TEMPLETON /4.5/ increased the accuracy of the frequency determination in Cu, Ag, and Au to 1 part in 10^6 using an NMR probe for the field measurements. An accurate and detailed determination of the Fermi surface of the noble metals which is used extensively in the following sections is the work done by HALSE /4.6/. In his paper an analytical expression for the form of the Fermi surface is given which is based on a symmetrized Fourier series. Since the Fermi surface is a periodic function in $\underline{k}$-space, it can be expanded in a Fourier series. In addition, the Fourier series must be invariant against the operation of the translational and point groups of the f.c.c. lattice of the noble metals. It turned out that an ansatz with five coefficients of the form (4.1) for the Fermi surface can describe the radii to 1 part in 10^3.

$$
\begin{aligned}
0 &= F(\underline{k}) \\
&= - C_{000} + (3 - \sum \cos \frac{a}{2}k_x \cos \frac{a}{2}k_y) \\
&\quad + C_{200}(3 - \sum \cos ak_x) \\
&\quad + C_{211}(3 - \sum \cos ak_x \cos \frac{a}{2}k_y \cos \frac{a}{2}k_z) \\
&\quad + C_{220}(3 - \sum \cos ak_x \cos ak_y) \\
&\quad + C_{310}(6 - \sum \cos \frac{3a}{2}k_x \cos \frac{a}{2}k_y - \sum \cos \frac{a}{2}k_x \cos \frac{3a}{2}k_y) \ .
\end{aligned}
\tag{4.1}
$$

The sums in (4.1) denote cyclic interchange of x, y, and z. The values of the lattice parameters a and of the coefficients $C_{\ell mn}$ used by HALSE are given in Table 4.1.

Table 4.1. Lattice parameters a and coefficients $C_{\ell mn}$ of the Fermi surface description (4.1) according to HALSE /4.6/

	a at 0 K
Cu	3.6030 ± 0.0004 Å
Ag	4.0692 ± 0.0008 Å
Au	4.0652 ± 0.0004 Å

	C_{000}	C_{200}	C_{211}	C_{220}	C_{310}
Cu5	1.69167	0.00693	-0.42501	-0.01679	-0.03772
Ag5	-0.89789	-0.12030	-0.90187	-0.14086	-0.09483
Au5	-2.26213	-0.16635	-1.25516	-0.09914	-0.12704

This representation of the Fermi surfaces of Cu, Ag, and Au has been used below in the determination of Fermi velocities from cyclotron masses and of scattering rates from Dingle temperatures. Fig. 4.3 shows the cross sections of the Fermi surfaces in the planes {100} and {110} drawn from the data published by HALSE. Silver shows the weakest and gold the strongest anisotropy of the noble metals.

Besides the experimental investigations, band structure calculations have increased the understanding of the Fermi surface of the noble metals. There exist a great number of calculations for the noble metals. Copper is the model substance

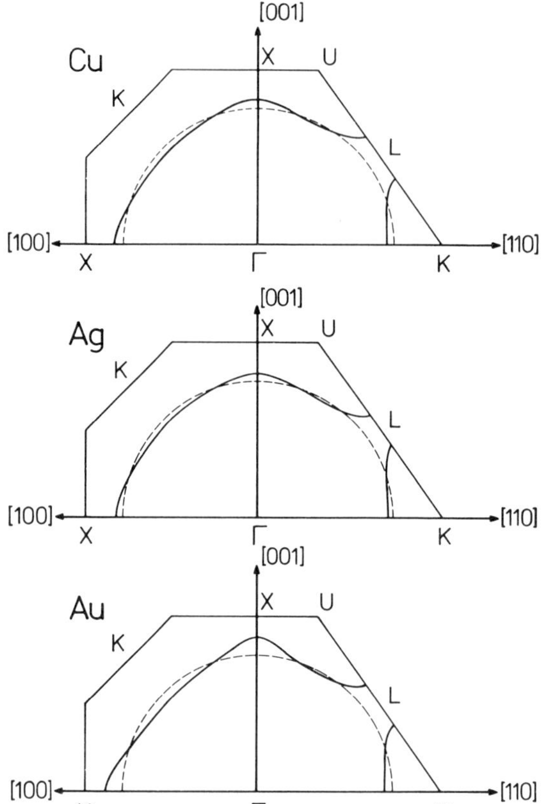

Fig. 4.3. Anisotropy of the Fermi surface of Cu, Ag, and Au in the planes {100} and {110} according to HALSE /4.6/

of a d-band metal for which many band structure techniques have been tested. The most usual ones are the augmented plane wave method (APW) /4.7/ and the Green's function method (KKR) /4.8/. The main problem in ab initio calculations is the construction of an electron potential which takes into account the Coulomb and the exchange interactions /4.9-13/. Even if these calculations are not able to give the Fermi surface with the same accuracy with which they can be determined experimentally, they show in a simple way the origin of the anisotropy sketched in Fig. 4.3. As described in SEGALL's paper for Cu /4.10/, the hybridization of the s-, p-, and d-bands is responsible for the deviations of the Fermi surface from a sphere. In the directions <100> and <111> the s- and p-bands hybridize. In the center Γ of the Brillouin zone the wave functions have pure s-character. At the boundary of the Brillouin zone (points X and L in Fig. 4.3), the wave functions have pure p-character. On the other hand the wave functions have s-character all along the direction <110> from Γ to the boundary point K. The s-, p-bands hybridize, in addition, with the d-bands which are completely occupied in the noble metals but which lie only a few eV below the Fermi level. The hybridization of the sp- and d-bands increases the energy of the states near the Fermi level (antibonding hybrids). Because the d-orbitals do not have lobes along the direction <111> there is no sp-d hybridization along these directions in contrast to all other directions. This implies that the states along <111> are not energetically enhanced. Consequently, they will be occupied up to higher $\underline{k}$-values compared to the other directions. That is the reason for the protrusions along the directions <111> which, due to the contact with the Brillouin zone, give rise to the necks. The sp-d hybridization is not equally strong in the different belly regions of the Fermi surface. Again for symmetry reasons the energy of the states <110> on the Fermi surface is enhanced more than the energy at the points <100>. For this reason the Fermi surface of the noble metals is bulged inwards at the points <110> and outwards at the points <100> (Fig. 4.3). The argument given here shows that general symmetry arguments can explain essential features of the Fermi surface of the noble metals.

Equation (4.1) is a phenomenological ansatz for the Fermi surface. The coefficients $C_{\ell mn}$ have no physical meaning. A theoretically more satisfactory parameterization of the Fermi surface can be given by means of a KKR-band structure calculation. For that purpose, certain parameters like the Fermi energy and the phase shifts $\eta_\ell^h(E_F)$ of the potential are chosen in such a way that the band structure reproduces the measured Fermi surface in an optimal way. Such a set of parameters for the noble metals has been calculated by LEE et al. /4.14/ and is given in Table 4.2.

Table 4.2. Fermi energy and phase shifts from a nonrelativistic band structure calculation for the noble metals according to LEE et al. /4.14/. The phase shifts for $\ell \geqslant 3$ are neglected

	$E_F[R_y]$	η_0^h	η_1^h	η_2^h
Cu	0.55	0.0755	0.1298	-0.1186
Ag	0.41	0.2097	0.1188	-0.1019
Au	0.53	0.2496	0.0632	-0.2426

These data will be used in Section 6 for the analysis of Dingle temperatures. Since the Fermi energy is not determined by the dHvA effect, it must be considered as a free parameter.

5. Cyclotron Masses and Fermi Velocities of the Noble Metals

5.1 Cyclotron Masses of Cu, Ag, and Au

According to Section 2.3.2 the cyclotron masses m_c^* can be determined from the temperature dependence of the dHvA amplitudes. In an extensive investigation we have measured a great number of cyclotron masses in Cu, Ag, and Au /5.1/. Fig. 5.1 shows the temperature dependence of the amplitudes for one orbit in the three metals. The

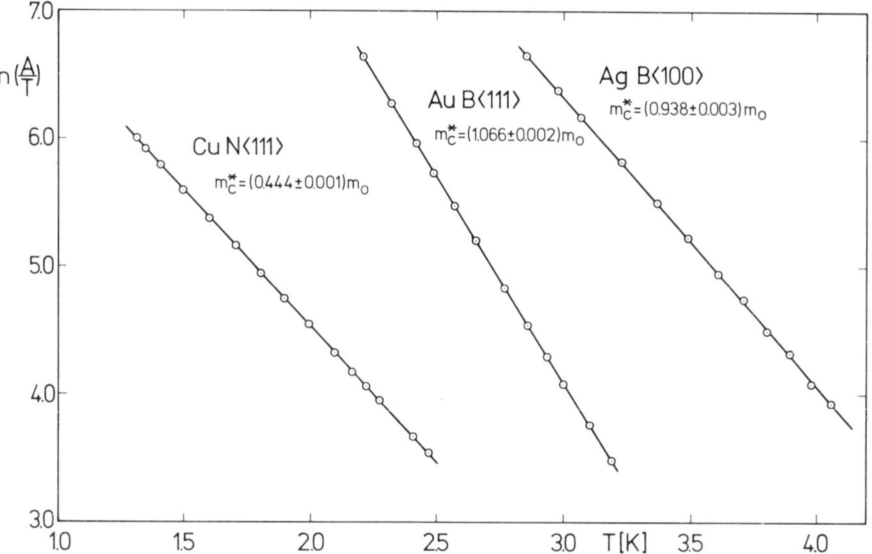

Fig. 5.1. Cyclotron masses for three extremal cross sections in Cu, Ag, and Au determined from the temperature dependence of the dHvA amplitudes. The slope of the lines is $-bm_c^*/m_0H$

temperature T has been determined from the vapor pressure of pumped ^{4}He /5.2/ by means of a capacitance manometer. The accuracy of the temperature reading is 1 mK below and a few mK above the λ-point $T_\lambda = 2.172$ K. Special attention has been given to the influence of systematic errors in the amplitude measurements. The greatest problems arose from the skin effect of the modulation field in the high purity samples. The magnetoresistance increases the skin depth to such a degree that a homogeneous penetration is guaranteed for all orbits, except for the orbits B<100> and D<110>, where it has a minimum /5.3/. Here the skin effect is still disturbing at 30 Hz. For noise reasons it was not possible to choose modulation frequencies ap-

Table 5.1. Cyclotron masses in copper. The experimental data have been determined from the temperature dependence of the dHvA amplitudes. The masses characterized by $C^{\pm}_{\ell mn}$ have been calculated from the energy surfaces adjacent to the Fermi surface and differing in energy by $\pm 6 \cdot 10^{-4}$ E_F from E_F

Cu	m_c^*/m_o		Cu	m_c^*/m_o	
	Experiment	$C^{\pm}_{\ell mn}$		Experiment	$C^{\pm}_{\ell mn}$
{110}			{110}		
B<100>		1.341	N75	0.648 ± 0.004	0.645
B0.5	1.343 ± 0.004	1.341	D85	1.288 ± 0.004	1.287
B10	1.315 ± 0.002	1.317	D89.5	1.260 ± 0.002	1.262
BTP	1.310 ± 0.003	1.309	D<110>		1.262
B50	1.388 ± 0.003	1.391			
B<111>	1.378 ± 0.003	1.375	{100}		
B65	1.431 ± 0.004	1.433	B7	1.326 ± 0.003	1.328
R<100>		1.306	BSP	1.320 ± 0.003	1.319
R0.5	1.307 ± 0.002	1.306	B18	1.327 ± 0.004	1.325
N<111>	0.444 ± 0.001	0.444	B29.2	1.468 ± 0.005	1.468
N65	0.478 ± 0.002	0.480	D40	1.309 ± 0.004	1.309

Table 5.2. Cyclotron masses in silver. See Table 5.1

Ag	m_c^*/m_o		Ag	m_c^*/m_o	
	Experiment	$C^{\pm}_{\ell mn}$		Experiment	$C^{\pm}_{\ell mn}$
{110}			{110}		
B<100>	0.938 ± 0.003	0.936	B65	0.954 ± 0.003	0.953
B7	0.923 ± 0.002	0.924	R<100>	1.044 ± 0.004	1.044
B8	0.919 ± 0.002	0.921	N<111>	0.365 ± 0.001	0.366
B10	0.916 ± 0.003	0.915	N60	0.375 ± 0.001	0.374
B12	0.911 ± 0.002	0.909	D85	1.030 ± 0.002	1.031
BTP	0.904 ± 0.003	0.903	D<110>	1.001 ± 0.002	1.000
B24	0.934 ± 0.002	0.934	{100}		
B50	0.928 ± 0.002	0.929	B4	0.933 ± 0.002	0.932
B<111>	0.923 ± 0.003	0.920	B7	0.923 ± 0.002	0.925
B60	0.927 ± 0.003	0.929	BSP	0.912 ± 0.002	0.912

Table 5.3. Cyclotron masses in gold. See Table 5.1

Au	m_c^*/m_o Experiment	$C^{\pm}_{\ell mn}$	Au	m_c^*/m_o Experiment	$C^{\pm}_{\ell mn}$
{110}			R1	1.014 ± 0.004	1.014
B<100>		1.142	N<111>	0.280 ± 0.001	0.281
B1	1.140 ± 0.004	1.141	N60	0.286 ± 0.001	0.286
B5	1.121 ± 0.003	1.122	D85	1.003 ± 0.003	1.003
B10	1.084 ± 0.002	1.083	D89.5	0.983 ± 0.002	0.983
B15	1.051 ± 0.003	1.050	D<110>		0.983
B23	1.052 ± 0.002	1.053	{100}		
B50	1.074 ± 0.005	1.074	B8	1.107 ± 0.002	1.106
B<111>	1.066 ± 0.002	1.065	BSP	1.067 ± 0.002	1.067
B60	1.071 ± 0.002	1.072	B25	1.073 ± 0.003	1.072
R<100>		1.014	D40	1.018 ± 0.003	1.018

preciably smaller than 30 Hz. Thus the masses have been measured for those orbits which are 0.5^0 away from <100> and <110> instead of the orbits B<100> and D<110> themselves. There the magnetoresistance is already large enough so that the skin effect can be neglected. In silver, a sample containing about 80 ppm of vacancies has been investigated. The vacancy induced resistivity made measurements also possible for the orbits B<100> and D<110>. Details of the temperature and amplitude measurements are described in /5.1/.

The Tables 5.1, 5.2, and 5.3 contain the values of the cyclotron masses for Cu, Ag, and Au measured in the planes {100} and {110}. B, R, N, and D denote belly, four-cornered rosette, neck, and dogsbone orbits, respectively. The numbers without brackets after the symbol give the angle in degrees by which the magnetic field is tilted against the crystallographic axis [001]. The positions of the planes {100} and {110} and of the angles θ and Φ are given in Fig. 5.2 to illustrate the orien-

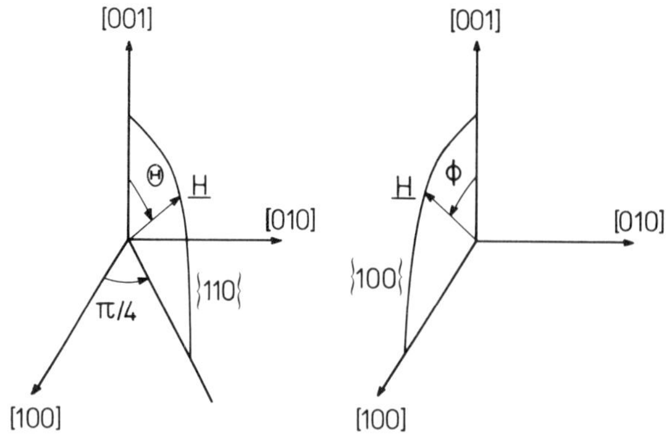

Fig. 5.2. The orientation of the magnetic field $\underline{H}$ relative to the crystal axes k_x, k_y, and k_z in the planes {100} and {110} is given by the angles Φ and Θ

tation of the different extremal cross sections. BTP and BSP are two stationary belly orbits with the orientations given in Table 5.4.

Table 5.4. Position of the belly turning point and belly saddle point orbits in the planes {110} and {100}

	Cu	Ag	Au
BTP {110} Θ =	16.2 ± 0.2	18.4 ± 0.1	21.5 ± 0.2
BSP {100} Φ =	11.8 ± 0.1	13.6 ± 0.1	16.3 ± 0.1

The accuracy of the measured cyclotron masses is ± 0.3%. A comparison with previous mass data published in the literature is given in Section 5.3.

5.2 Determination of Energy Surfaces Adjacent to the Fermi Surface

According to (2.51) the cyclotron masses are orbital averages of the reciprocal Fermi velocities $v^*(\underline{k})$. In principle, another symmetrized Fourier series can be chosen which parameterizes the Fermi velocities, similar to that used by HALSE to parameterize the Fermi surface. But an ansatz with five coefficients for $v^*(\underline{k})$ produces poor correlation with the data although the same ansatz describes the Fermi surface quite well. The reason for this is the stronger anisotropy of the Fermi velocities. To keep the number of fit parameters as small as possible, the parametrization scheme proposed by HALSE has been adopted here. It consists of constructing energy surfaces adjacent to the Fermi surface which differ in energy from the Fermi energy E_F by $\delta E/E_F = \pm 6 \cdot 10^{-4}$ (Fig. 5.3). These two surfaces are described again by a symme-

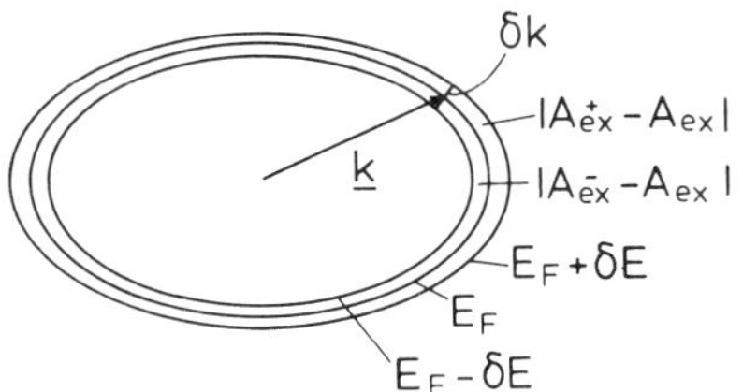

Fig. 5.3. The distance δk of the two surfaces $E_F \pm \delta E$ in $\underline{k}$ is inversely proportional to the Fermi velocity $v^*(\underline{k})$

trized Fourier series with coefficients $C^{\pm}_{\ell mn}$. If A^+_{ex} , A^-_{ex} , and A_{ex} are the areas of the extremal cross sections for a given field directions on the surfaces $E_F + \delta E$, $E_F - \delta E$, and E_F then

$$m_c^*/m_o = |A_{ex}^{\pm} - A_{ex}| E_F / \pi k_s^2 \ \delta E \ . \qquad (5.1)$$

The coefficients $C_{\ell mn}^{\pm}$ are determined in such a way that the $A_{ex}^{\pm}$ fit the measured masses according to (5.1). This has been done by means of a nonlinear least squares fitting program (VA05AD Harwell Subroutine Library). All masses from the Tables 5.1, 5.2 , and 5.3 have been used in the fit. The values of the fitted parameters $C_{\ell mn}^{\pm}$ are given in Table 5.5. The masses, which have been calculated from the coefficients

Table 5.5. Coefficients $C_{\ell mn}^{\pm}$ of the Fourier series which describe the energy surfaces adjacent to the Fermi surface. The energy difference between the surfaces $E_F \pm \delta E$ and E_F is $\pm 6 \cdot 10^{-4}$ E_F. The coefficients $C_{\ell mn}$ for the Fermi surface are those by HALSE /4.6/

	C_{000}	C_{200}	C_{211}	C_{220}	C_{310}
Cu5	1.69167	0.00693	-0.42501	-0.01679	-0.03772
Cu5+	1.705925	0.007730	-0.422119	-0.016660	-0.037677
Cu5-	1.680147	0.006285	-0.427345	-0.016834	-0.037728
Ag5	-0.89789	-0.12030	-0.90187	-0.14086	-0.09483
Ag5+	-0.872823	-0.118934	-0.896995	-0.140175	-0.094590
Ag5-	-0.911864	-0.121070	-0.904573	-0.141142	-0.094881
Au5	-2.26213	-0.16635	-1.25516	-0.09914	-0.12704
Au5+	-2.385641	-0.172520	-1.280091	-0.102875	-0.129762
Au5-	-2.141185	-0.160308	-1.230744	-0.095484	-0.124378

$C_{\ell mn}^{+}$ and $C_{\ell mn}^{-}$, are compared to the experimental data in Tables 5.1 , 5.2 , and 5.3. The fitted masses agree within the accuracy of the data with the measured ones.

5.3 Angular Dependence of the Cyclotron Masses in Cu, Ag, and Au

Once the coefficients $C_{\ell mn}^{\pm}$ are known, the cyclotron masses can be determined for any orientation of the crystal in the field using (5.1). In Table 5.6 a number of masses are quoted which have been calculated in this way. Using these data a plot of the cyclotron masses in the planes {100} and {110} has been drawn in Fig. 5.4 for Cu, Ag, and Au. For comparison, some data from the literature are shown as well. These are the cyclotron resonance data by KOCH et al. /5.4/ for Cu, by HOWARD /5.5/ for Ag, and by LANGENBERG et al. /5.6/ for Au; and dHvA data by COLERIDGE et al /5.7/

Table 5.6. Cyclotron masses m_c^*/m_o for Cu, Ag, and Au calculated from the energy surfaces $E_F \pm \delta E$. The angles Θ and Φ are those of Fig. 5.2

	Cu	Ag	Au		Cu	Ag	Au
BΘ = 0	1.341	0.936	1.142	DΘ = 90	1.262	1.000	0.983
5	1.333	0.930	1.122	88	1.266	1.005	0.986
10	1.317	0.915	1.083	85	1.287	1.031	1.003
15	1.308	0.903	1.050	83	1.313	1.066	1.025
20	1.325	0.907	1.038	80	1.377	1.178	1.078
25	1.481	0.949	1.096	78	1.446	-	1.137
26	1.714	0.970	1.207	DΦ = 43	1.269	1.009	0.988
27.5	-	1.035	1.426	40	1.309	1.065	1.018
45	1.664	0.977	1.165	37.5	1.381	1.218	1.075
47	1.441	0.949	1.098	35	1.541	-	1.217
50	1.391	0.929	1.074	RΘ = 0	1.306	1.044	1.014
54.74	1.375	0.920	1.065	5	1.417	1.218	1.102
60	1.388	0.929	1.072	NΘ = 30	0.952	0.849	0.576
65	1.433	0.953	1.095	32	0.774	0.662	0.478
70	1.558	1.004	1.155	35	0.638	0.534	0.398
73	2.013	1.061	1.286	40	0.529	0.438	0.333
BΦ = 5	1.333	0.930	1.124	45	0.476	0.393	0.300
10	1.322	0.918	1.094	50	0.451	0.372	0.285
15	1.319	0.911	1.071	54.74	0.444	0.366	0.281
20	1.333	0.917	1.061	60	0.453	0.374	0.286
25	1.377	0.941	1.072	65	0.380	0.397	0.303
30	1.500	0.998	1.130	70	0.534	0.444	0.336
32	1.653	1.041	1.199	75	0.645	0.546	0.404
33	1.956	1.072	1.282	77	0.722	0.624	0.450

for Cu, by JOSEPH et al. /5.8/ for Ag, and by BOSACCHI et al. /4.4/ for Au. In Section 2.2 it has been pointed out that the same enhanced mass is measured in the dHvA effect and in cyclotron resonance. Mass measurements by the dHvA effect have two advantages compared to those by cyclotron resonance. First, masses deduced from the Azbel-Kaner theory of the cyclotron resonance are only correct when the magnetic field is parallel to the crystal surface. Even small tilts such as those produced by surface roughness can seriously affect the data /5.4/. This problem does not arise in the dHvA effect. Secondly, the dHvA effect itself can be used to orient the crystals in the field. Due to very careful temperature and amplitude measurements the masses quoted in Tables 5.1, 5.2, and 5.3 are probably the most reliable set of data for the noble metals available at the present time.

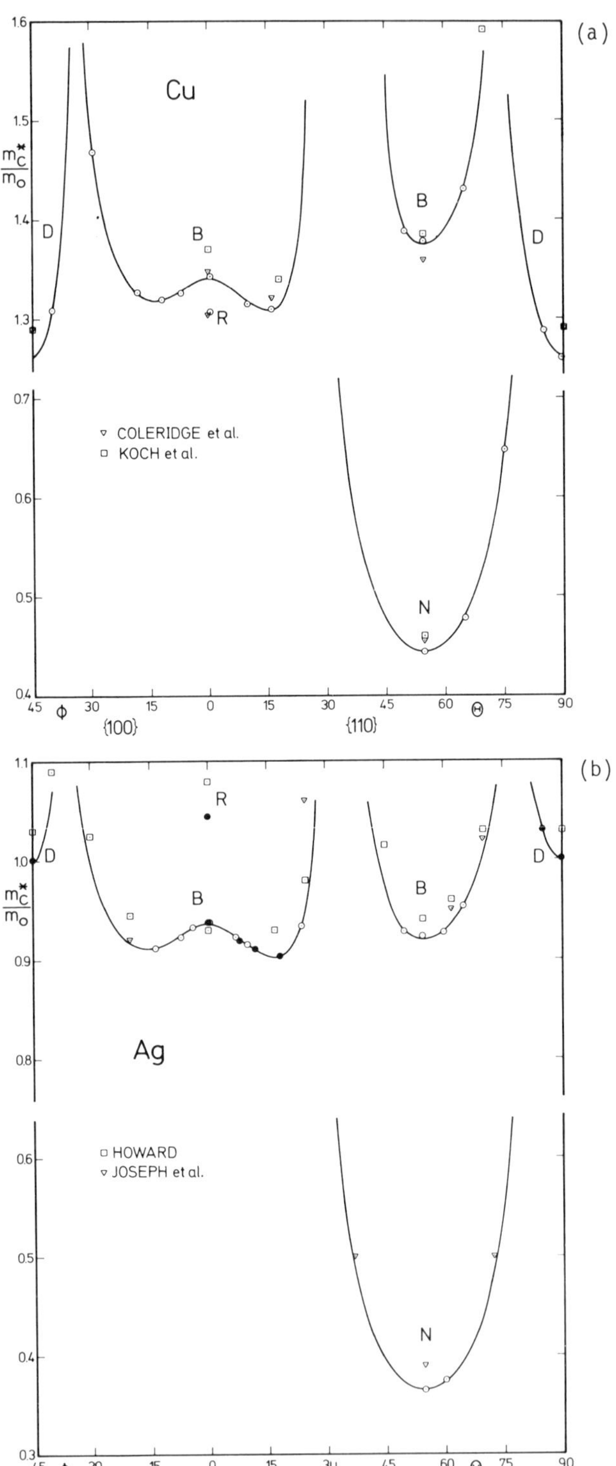

Fig. 5.4a-c. Angular dependence of the cyclotron masses in Cu, Ag, and Au. The present results are indicated by circles. For comparison some cyclotron resonance data (squares) and some dHvA data (triangles) are shown as well

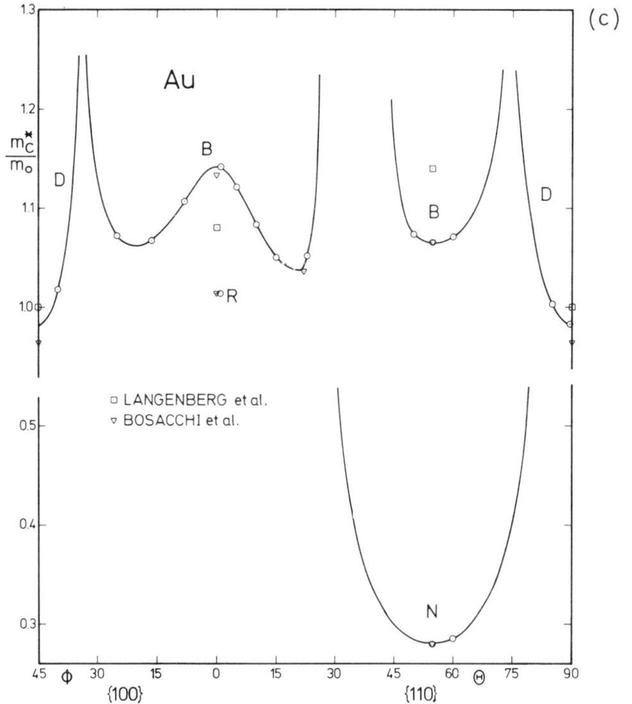

Fig. 5.4c

5.4 Fermi Velocities in the Noble Metals

If δk is the distance of the two surfaces $E_F + \delta E$ and $E_F - \delta E$ at the point $\underline{k}$ of the Fermi surface (Fig. 5.3), the Fermi velocity in $\underline{k}$ is given by

$$v^*(\underline{k})/v_s = 6\cdot 10^{-4}\ k_s/\delta k \tag{5.2}$$

with

$$v_s = \hbar k_s/m_o \tag{5.3}$$

$$k_s = (12\pi^2)^{1/3} 1/a\ . \tag{5.4}$$

k_s and v_s are the radius and the velocity for the free electron sphere of energy E_F. For an energy difference $\delta E/E_F = 6\cdot 10^{-4}$, the gradient $\hbar v^* = |\partial E/\partial \underline{k}|$ can be replaced in (5.2) by the difference quotient $\delta E/\delta k$. The Fermi velocities calculated in this way are given in the Tables 5.7 , 5.8, and 5.9 for Cu, Ag, and Au. ϕ and θ are the usual polar coordinates with the pole at [001]. $\phi = \pi/2$ and $\theta = 0$ are the coordinates of the point [100]. Since the necks are circular to 1 part in 10^{-4}, the Fermi velocity at the neck periphery is

$$v_N^*/v_s = (A_{exN}/\pi k_s^2)^{1/2}(m_o/m_{cN}^*)\ . \tag{5.5}$$

Table 5.7. Fermi velocities $v^*(\underline{k})/v_s$ in Cu deduced from the cyclotron masses in Table 5.1 (v_s = $1.5779\cdot10^8$ cm s^{-1})

	ϕ = 0	5	10	15	20	25	30	35	40	45
θ = 0	0.684	0.684	0.684	0.684	0.684	0.684	0.684	0.684	0.684	0.684
5	0.727	0.727	0.727	0.727	0.727	0.727	0.727	0.727	0.727	0.727
10	0.790	0.790	0.790	0.790	0.790	0.790	0.790	0.790	0.790	0.790
15	0.813	0.813	0.813	0.813	0.814	0.814	0.814	0.815	0.815	0.815
20	0.800	0.800	0.801	0.803	0.804	0.806	0.808	0.809	0.810	0.810
25	0.772	0.773	0.775	0.778	0.782	0.785	0.788	0.791	0.792	0.793
30	0.743	0.744	0.747	0.752	0.757	0.763	0.767	0.770	0.771	0.772
35	0.718	0.720	0.724	0.730	0.736	0.742	0.744	0.744	0.741	0.740
40	0.702	0.704	0.708	0.715	0.720	0.721	0.714	0.696	0.674	0.664
45	0.696	0.698	0.703	0.709	0.711	0.700	0.664	0.585	0.477	0.428
50	0.702	0.704	0.708	0.713	0.711	0.685	0.600	-	-	-
55	0.718	0.720	0.724	0.727	0.722	0.689	0.581	-	-	-
60	0.743	0.744	0.747	0.748	0.741	0.714	0.634	0.439	-	-
65	0.772	0.773	0.773	0.771	0.763	0.742	0.701	0.625	0.522	0.464
70	0.800	0.800	0.798	0.792	0.781	0.763	0.739	0.709	0.680	0.667
75	0.813	0.814	0.813	0.807	0.793	0.773	0.752	0.731	0.715	0.708
80	0.790	0.799	0.813	0.814	0.798	0.775	0.750	0.728	0.713	0.708
85	0.727	0.756	0.799	0.814	0.800	0.773	0.745	0.721	0.705	0.700
90	0.684	0.727	0.790	0.813	0.800	0.772	0.743	0.718	0.702	0.696

The value of v_N^*/v_s is (0.425 ± 0.001) for Cu, (0.371 ± 0.001) for Ag, and (0.638 ± 0.002) for Au. The accuracy of the velocities is better than 1%, especially at the neck where it is within 0.3%. These data are significantly more accurate than the data published by HALSE which were deduced from older mass data /4.6/. They had an accuracy of 3% for Cu and Ag and of 10% for Au. The anisotropy of the Fermi velocities $v^*(\underline{k})/v_s$ along some symmetry directions is shown in Fig. 5.5. The qualitative behavior of the velocities is the same in the three metals.

Fermi velocities can also be deduced from Landau surface states /5.9/. If the Fermi surface is known, the velocities can be determined from the position of the

Table 5.8. Fermi velocities $v^*(\underline{k})/v_s$ in Ag deduced from the cyclotron masses in Table 5.2 ($v_s = 1.3971 \cdot 10^8$ cm s^{-1})

	ϕ = 0	5	10	15	20	25	30	35	40	45
θ = 0	0.927	0.927	0.927	0.927	0.927	0.927	0.927	0.927	0.927	0.927
5	0.976	0.976	0.976	0.976	0.976	0.976	0.976	0.976	0.976	0.976
10	1.067	1.067	1.067	1.067	1.068	1.068	1.068	1.069	1.069	1.069
15	1.129	1.129	1.130	1.131	1.133	1.134	1.136	1.137	1.138	1.138
20	1.145	1.146	1.148	1.151	1.155	1.159	1.162	1.165	1.167	1.168
25	1.132	1.133	1.136	1.141	1.147	1.153	1.159	1.163	1.166	1.167
30	1.106	1.108	1.112	1.118	1.125	1.131	1.136	1.139	1.141	1.141
35	1.081	1.082	1.087	1.092	1.097	1.100	1.098	1.092	1.087	1.084
40	1.063	1.064	1.068	1.071	1.071	1.061	1.039	1.007	0.975	0.962
45	1.056	1.058	1.061	1.061	1.051	1.021	0.957	0.855	0.740	0.683
50	1.063	1.064	1.067	1.064	1.046	0.992	0.872	0.638	-	-
55	1.081	1.082	1.085	1.083	1.061	0.997	0.850	0.529	-	-
60	1.106	1.108	1.112	1.111	1.092	1.037	0.919	0.699	-	-
65	1.132	1.135	1.140	1.141	1.127	1.088	1.016	0.907	0.785	0.725
70	1.145	1.150	1.159	1.164	1.154	1.127	1.084	1.031	0.982	0.962
75	1.129	1.138	1.158	1.170	1.165	1.144	1.113	1.080	1.053	1.043
80	1.067	1.087	1.129	1.158	1.161	1.143	1.116	1.089	1.068	1.061
85	0.976	1.014	1.087	1.138	1.151	1.136	1.110	1.084	1.065	1.059
90	0.927	0.976	1.067	1.129	1.145	1.132	1.106	1.081	1.063	1.056

resonance in the surface impedance. Whereas in the dHvA effect the measured masses are averages of the reciprocal Fermi velocity averaged over the full orbit of an extremal cross section, the velocities are averaged only over a strip of 5 - 10^0 in the Landau surface states. This is an advantage for the determination of the velocities. On the other hand the stringent requirements concerning the purity and surface conditions of the sample limit this method to a small number of metals. Some Fermi velocities have been measured for Cu and Ag /5.9,10/. The measured values are shown in Fig. 5.5. For Cu, the surface preparation is well established. Here the agreement with our data is satisfactory. In Ag the surface preparation creates problems /5.10/

Table 5.9. Fermi velocities $v^*(\underline{k})/v_s$ in Au deduced from the cyclotron masses in Table 5.3 ($v_s = 1.3985 \cdot 10^8$ cm s^{-1})

	$\phi = 0$	5	10	15	20	25	30	35	40	45
$\theta = 0$	0.736	0.736	0.736	0.736	0.736	0.736	0.736	0.736	0.736	0.736
5	0.891	0.891	0.891	0.891	0.891	0.891	0.891	0.891	0.891	0.891
10	1.014	1.014	1.015	1.015	1.017	1.018	1.019	1.020	1.021	1.021
15	1.028	1.029	1.031	1.035	1.040	1.045	1.050	1.054	1.057	1.058
20	0.997	0.999	1.004	1.013	1.023	1.034	1.045	1.054	1.061	1.063
25	0.952	0.955	0.963	0.976	0.993	1.012	1.030	1.046	1.056	1.059
30	0.905	0.909	0.920	0.938	0.961	0.987	1.012	1.034	1.048	1.053
35	0.866	0.870	0.884	0.906	0.934	0.964	0.993	1.016	1.030	1.035
40	0.839	0.845	0.860	0.885	0.915	0.946	0.970	0.978	0.974	0.969
45	0.830	0.836	0.853	0.879	0.910	0.934	0.931	0.872	0.757	0.690
50	0.839	0.846	0.865	0.893	0.922	0.934	0.878	0.640	-	-
55	0.866	0.873	0.894	0.924	0.953	0.956	0.868	-	-	-
60	0.905	0.913	0.936	0.967	0.994	0.998	0.935	0.708	-	-
65	0.952	0.960	0.983	1.012	1.033	1.035	1.002	0.923	0.806	0.737
70	0.997	1.006	1.026	1.047	1.055	1.045	1.017	0.979	0.942	0.925
75	1.028	1.036	1.054	1.062	1.053	1.026	0.990	0.954	0.927	0.916
80	1.014	1.031	1.055	1.055	1.031	0.993	0.950	0.911	0.884	0.874
85	0.891	0.961	1.031	1.037	1.007	0.963	0.917	0.878	0.851	0.842
90	0.736	0.891	1.014	1.028	0.997	0.952	0.905	0.866	0.839	0.830

and discrepancies of up to 13% occur with our data. This value is much greater than the errors quoted. In Au it has not yet been possible to observe magnetic surface states.

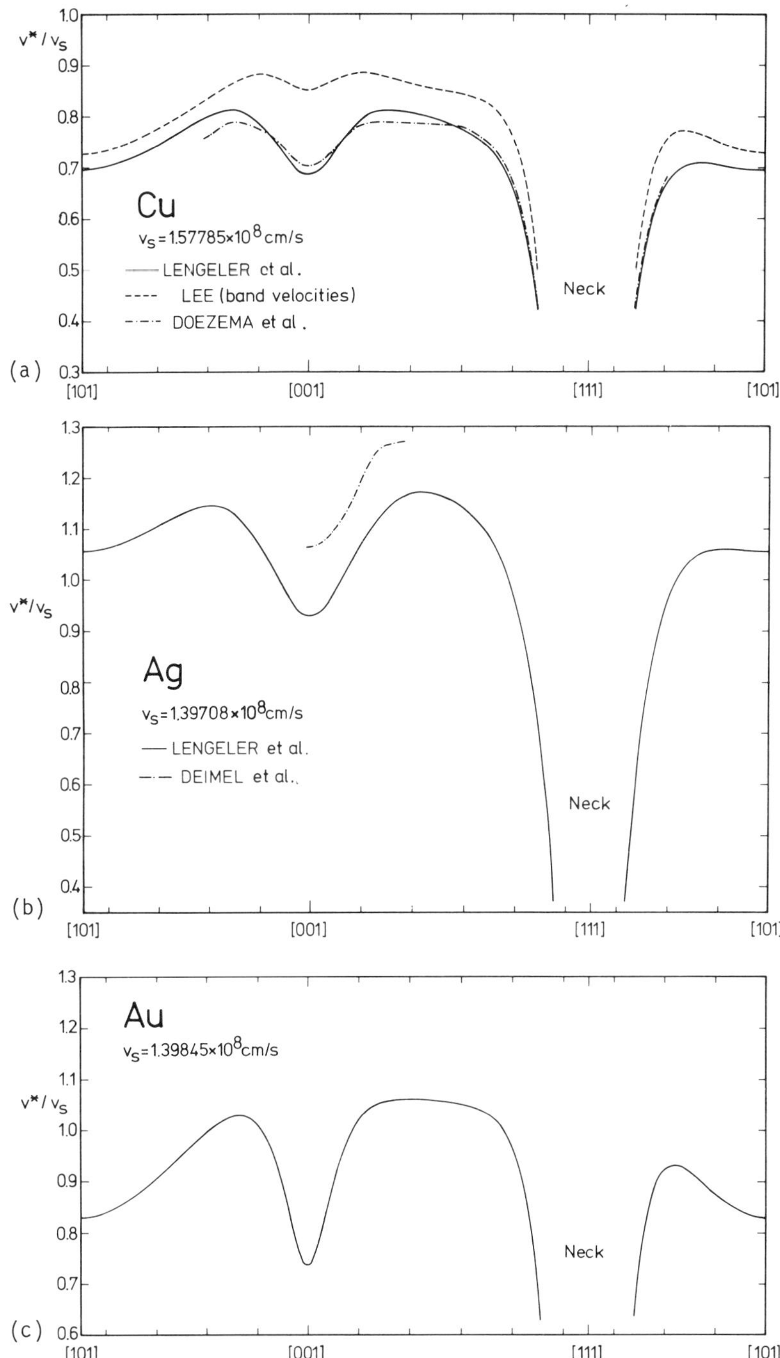

Fig. 5.5a - c. Fermi velocities in Cu, Ag, and Au along some high symmetry directions. Our data are given by the solid lines. Data deduced from magnetic surface states are dot-dashed. Band velocities by LEE /5.14/ are dashed

5.5 Electron-Phonon Coupling Constant $\lambda(\underline{k})$ in Cu

Cyclotron masses and Fermi velocities which have been determined from band structure calculations are not renormalized by electron-phonon interaction. A comparison with dHvA data can therefore give values of $\lambda(\underline{k})$ and $\langle\lambda\rangle$.

$$1 + \lambda(\underline{k}) = v(\underline{k})/v^*(\underline{k}) \qquad (5.6)$$

$$1 + \langle\lambda\rangle = m_c^*/m_c \ . \qquad (5.7)$$

As explained in Section 4, there are no potentials available which allow one to calculate the Fermi surface and the Fermi velocities with the same accuracy with which they can be measured. Calculations of the velocities have been carried out by LEWIS et al. /5.11/, O'SULLIVAN et al. /5.12/, and JANAK et al. /5.13/. The most accurate and detailed calculations have been made for Cu by LEE /5.14/. Table 5.10 gives some of the calculated masses and the corresponding constants $1+\langle\lambda\rangle$ obtained

Table 5.10. Measured and calculated /5.14/ cyclotron masses in Cu and corresponding coupling constants $1+\langle\lambda\rangle$ for some orbits in the plane {110}

Cu	m_c/m_o /5.14/	m_c^*/m_o Table 5.6	$1+\langle\lambda\rangle$
B<100>	1.238	1.341	1.083
B8	1.229	1.323	1.076
B16	1.223	1.308	1.070
B24	1.296	1.420	1.096
B<111>	1.277	1.375	1.077
R<100>	1.183	1.306	1.104
N30	0.827	0.952	1.151
N40	0.448	0.529	1.181
N50	0.380	0.451	1.187
N<111>	0.374	0.444	1.187
N60	0.383	0.453	1.183
N70	0.452	0.534	1.181
D<110>	1.126	1.262	1.121

by comparison with the present dHvA cyclotron masses. The band velocities calculated by LEE and the electron-phonon coupling constants $1+\lambda(\underline{k})$ deduced by comparison with the dHvA data are given in the Figs. 5.5 and 5.6. $\lambda(\underline{k})$ is strongly anisotropic and has maxima at the necks and at the belly points <100>. These are the regions where

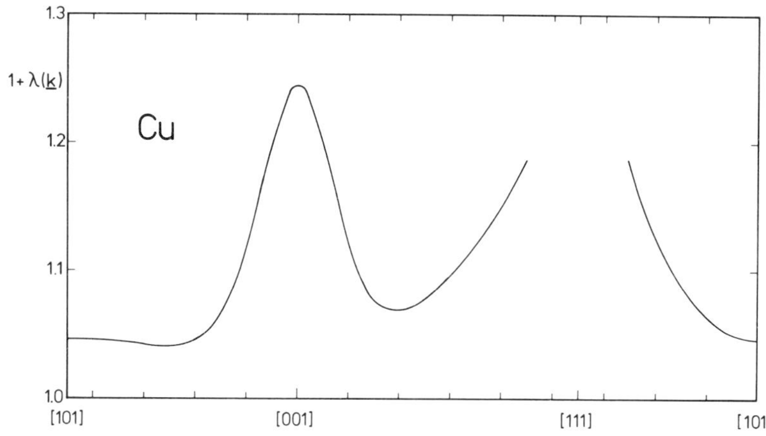

Fig. 5.6. Local values of the electron-phonon coupling constants $1+\lambda(\underline{k})$ in Cu calculated from the Fermi velocities $v^*(\underline{k})$ and the band velocities $v(\underline{k})$ of Fig. 5.5

the Fermi surface deviates the most from the free electron sphere. To our knowledge there are no experimental values of the $\underline{k}$-dependence of λ, but there are mean values $\bar{\lambda}$ of λ averaged over the whole Fermi surface as shown in Table 5.11. The mean value obtained from the present $\lambda(\underline{k})$ is 0.11, which is somewhat lower than the values from the table. In view of the limited accuracy of the band velocities on which the values $\lambda(\underline{k})$ are based, this discrepancy is not significant.

Table 5.11. Values of the electron-phonon coupling $\bar{\lambda}$ averaged over the Fermi surface of Cu

$\bar{\lambda}$	Experiment	$\bar{\lambda}$	Theory
0.16	/5.15/	0.15	/5.17/
0.13 ± 0.03	/5.16/	0.15 ± 0.02	/5.18/
0.11	this work	0.12 ± 0.02	/5.19/
		0.14	/5.20/

5.6 Coefficient of the Electronic Specific Heat for Cu, Ag, and Au

At low temperatures the electronic contribution C_{el} to the specific heat increases linearly with T. The proportionality constant γ^* is given by

$$\gamma^* = \frac{1}{3}\pi^2 k_B^2 D^*(E_F) \ . \tag{5.8}$$

The density of states $D^*(E_F)$ at the Fermi level is enhanced by electron-phonon interaction. If δV_k is the volume between the two surfaces $E_F \pm \delta E$ characterized by the coefficients $C^{\pm}_{\ell mn}$

$$D^*(E_F) = (2\pi)^{-3}\ \delta V_k/\delta E \tag{5.9}$$

and thus

$$\gamma^*/\gamma_s = 10^4\ \delta V_k/18V_s \tag{5.10}$$

with

$$\gamma_s = k_B^2 m_o k_s/3\hbar^2 \tag{5.11}$$

$$k_s = (12\pi^2)^{1/3}\, 1/a \tag{5.12}$$

$$V_s = 4\pi k_s^3/3\ . \tag{5.13}$$

k_s and V_s are the radius and the volume of the free electron sphere for E_F. In Table 5.12 are given the coefficients γ^*/γ_s calculated from the present dHvA data. They agree very well with the values obtained by MARTIN from specific heat measurements /5.21/. The close agreement supports the reliability of the present cyclotron mass measurements. Earlier calculations of γ^*/γ_s by HALSE /4.6/ and by BOSACCHI et al. /4.4, 5.22/ are given for comparison in Table 5.12.

Table 5.12. Coefficients γ^*/γ_s of the electronic specific heat in free electron units γ_s for Cu, Ag, and Au. γ_s is 0.49954, 0.63718, and 0.63592 mJ/mole·K^2 for Cu, Ag, and Au

	γ^*/γ_s	
Cu	1.382 ± 0.010	this work
	1.383 ± 0.002	MARTIN /5.21/
	1.397 ± 0.016	BOSACCHI et al. /5.22/
	1.400	HALSE /4.6/
Ag	1.008 ± 0.007	this work
	1.004 ± 0.002	MARTIN /5.21/
	1.021	HALSE /4.6/
Au	1.074 ± 0.007	this work
	1.083 ± 0.002	MARTIN /5.21/
	1.077	BOSACCHI et al. /4.4/
	1.093	HALSE /4.6/

6. Dingle Temperatures and Scattering Rates of Conduction Electrons in the Noble Metals

6.1 Dingle Temperatures and the Lifetime of Electron States

The conduction electrons are scattered by impurities and intrinsic defects (vacancies, interstitials, dislocations, etc.). The resulting reduced lifetime of the electrons in a given state causes a broadening of the Landau levels, which is described by a Dingle temperature X^*. The product $m_c^* X^*$ can be determined from the field dependence of the dHvA amplitudes. If m_c^* is taken from the temperature dependence of the dHvA amplitudes, then X^* can be deduced from

$$m_c^* X^* = m_c X \;, \tag{6.1}$$

since this product is independent of the electron-phonon interaction. The Dingle temperature X^* is an average of the local scattering rates of the conduction electrons for a given extremal cross section

$$X^* = (\hbar/2\pi k_B) \; \langle 1/\tau^*(\underline{k}) \rangle \;. \tag{6.2}$$

The individual elements dk of the extremal cross section are weighted in (6.2) by the time spent in them by the $\underline{k}$-vector /6.1/ (Fig. 2.3).

$$X^* = (\hbar\omega_c^*/4\pi^2 k_B) \oint dt \; 1/\tau^*[\underline{k}(t)] \tag{6.3}$$

or

$$X^* m_c^* = (\hbar^2/4\pi^2 k_B) \oint dk/v^* \tau^*(\underline{k}) \tag{6.4}$$

$$= (\hbar^2/4\pi^2 k_B) \oint d\alpha \; k_\perp^2 [(\underline{v}^* \cdot \underline{k}_\perp) \tau^*(\underline{k})]^{-1} \;. \tag{6.5}$$

The weighting factor $k_\perp^2/(\underline{v}^* \cdot \underline{k}_\perp)$ of the scattering rate depends on the geometry of the extremal cross section and on the associated Fermi velocities. Just like $X^* m_c^*$ the product $v^* \tau^*$ in (6.5) is independent of the electron-phonon interaction

$$\tau^* v^* = \tau v \;. \tag{6.6}$$

The enhancement of the electron lifetime $\tau^*(\underline{k})$ by electron-phonon interaction

$$\tau^*(\underline{k}) = \tau(\underline{k}) \; [1 + \lambda(\underline{k})] \tag{6.7}$$

is a consequence of the scattering of the conduction electrons by virtual phonons which increase the inertia of the electrons and thereby reduce the scattering probability.

According to Bohr-Onsager quantization, the interference of the electron wave with itself on the cyclotron orbit brings about the redistribution of the electronic states on Landau cylinders in a magnetic field. Every scattering event which destroys the phase coherence of the wave produces a reduction of the dHvA amplitude. This reduction is particularly large for scattering angles θ

$$\theta = \pi/n = \pi H/F \tag{6.8}$$

where the phase of the wave is shifted by π. Since H/F is typically 10^{-3} to 10^{-4} in the noble metals, the phase coherence is destroyed for scattering angles even smaller than 0.1^{0}. For that reason, $\tau(\underline{k})$ can be considered as the real lifetime of an electron in state $\underline{k}$. Here it should be emphasized again that the scattering of the electrons at real phonons seems not to contribute to the reduction of $\tau(\underline{k})$ (see Sec. 2.2). If $P(\underline{k},\underline{k}')$ is the transition rate from a state $\underline{k}$ to a state $\underline{k}'$ then

$$1/\tau(\underline{k}) = (2\pi)^{-3} \int d^3\underline{k}' \; P(\underline{k},\underline{k}') \; . \tag{6.9}$$

According to Fermi's golden rule /6.2/ the transition rate is

$$P(\underline{k},\underline{k}') = (2\pi c_d/\hbar) \; |T_{\underline{k}\underline{k}'}|^2 \; \delta[E(\underline{k}) - E(\underline{k}')] \tag{6.10}$$

where c_d is the concentration of scattering centers in the lattice in the dilute limit. Thus

$$1/\tau(\underline{k}) = (2\pi)^{-3}(2\pi c_d/\hbar) \int_{FS} dS_{\underline{k}'} \; |T_{\underline{k}\underline{k}'}|^2/\hbar v(\underline{k}') \; . \tag{6.11}$$

The surface integral extends over the Fermi surface. Using the optical theorem, which expresses the conservation of particles,

$$(2\pi)^{-3} \int dS_{\underline{k}'} \; |T_{\underline{k}\underline{k}'}|^2/\hbar v(\underline{k}') = -\mathrm{Im}\{T_{\underline{k}\underline{k}}/\pi\} \tag{6.12}$$

(6.11) can be written

$$1/\tau(\underline{k}) = (2c_d/\hbar) \; \mathrm{Im}\{T_{\underline{k}\underline{k}}\} \; . \tag{6.13}$$

The scattering of the conduction electrons by magnetic impurities will not be considered in this article. This has been described in some detail by SHIBA /6.3/. The scattering of the electrons by nonmagnetic impurities in the noble metals is field independent. This is again a consequence of the high phase $2\pi F/H$ of the dHvA oscillations. In this case, the radius of a cyclotron orbit is appreciably larger than the linear dimensions of the scattering potentials. Hence the curvature of the orbit over the range of the potential can be neglected. In the quantum limit, which is not achieved here, the cyclotron orbit has atomic dimensions. Then this assumption does not hold any longer and $\tau(\underline{k})$ will be field dependent.

The lifetime $\tau(\underline{k})$ which enters the dHvA effect differs generally from the transport relaxation time $\tau_{tr}(\underline{k})$ which enters in transport coefficients such as the electrical resistivity. Small angle scattering processes do not contribute effectively to the resistance, because they scatter the electrons only slightly from the drift direction. This is taken into account by a weighting factor which can be often approximated by $(1-\cos\theta)$ where θ is the scattering angle. In this case, the lifetime becomes

$$1/\tau_{tr}(\underline{k}) = (2\pi)^{-3}\int d^3\underline{k}'\ P(\underline{k},\underline{k}')(1-\cos\theta)\ . \tag{6.14}$$

Electrons are scattered preferentially at small angles by the long-range strain field of dislocations. Hence, isolated dislocations are hardly visible in the electrical resistivity. On the other hand, they destroy the phase coherence of the electrons on a cyclotron orbit /6.4,5/. The dHvA effect is therefore a sensitive probe for dislocations in single crystals (Sec.3.4). There is another essential difference between the electron lifetime $\tau(\underline{k})$ and the transport relaxation time $\tau_{tr}(\underline{k})$. τ_{tr} is introduced as an approximate solution of the Boltzmann equation. Only for the trivial case of isotropic scattering and a spherical Fermi surface is τ_{tr} an exact solution of the Boltzmann equation. In general, the solution of this equation produces great difficulties. In contrast to this, the lifetime $\tau(\underline{k})$ which enters the dHvA effect is much more easily accessible from a theoretical standpoint. Furthermore, the transport coefficients contain averages of $\tau_{tr}(\underline{k})$ over the whole Fermi surface, whereas in the dHvA effect the average is only over an extremal cross section.

6.2 Anisotropy of the Scattering Rates in the Noble Metals

Dingle temperature determinations for dilute noble metal alloys and for noble metals containing intrinsic defects (dislocations, vacancies) have been reported by a number of authors. These include the investigations of POULSEN et al. on <u>Cu</u>Au, <u>Cu</u>Ge, and <u>Cu</u>Ni /6.6/; TEMPLETON et al. on <u>Cu</u>Al and <u>Cu</u>Ni /6.7/; WAMPLER et al. on <u>Cu</u>H /6.8/; BROWN et al. on <u>Ag</u>Au, <u>Ag</u>Cd, <u>Ag</u>Ge, and <u>Ag</u>Sn /6.9/; LOWNDES et al. on <u>Au</u>Ag, <u>Au</u>Cu, <u>Au</u>Zn /6.1/; and CHUNG et al. on <u>Au</u>Co /6.10/. Dislocations in Cu have been studied by COLERIDGE et al. /6.4/ and CHANG et al. /6.5/. Vacancies in Au have been investigated by LENGELER /6.11/ and CHANG et al. /6.12/. An example of the determination of Dingle temperatures from the field dependence of the dHvA amplitude is shown in Fig. 6.1. Although the error in the determination of the slope is typically 0.5%, the Dingle temperatures per at % defects are not nearly so accurately known. This is partly due to inaccuracies in the determination of the concentration of defects. The main contribution to the errors is due to the scattering of the electrons at defects (mainly dislocations) which are also present in the sample, often in unknown concentrations. A possibility to separate the contributions of the different defects is to measure the dependence of the Dingle temperatures on the defect con-

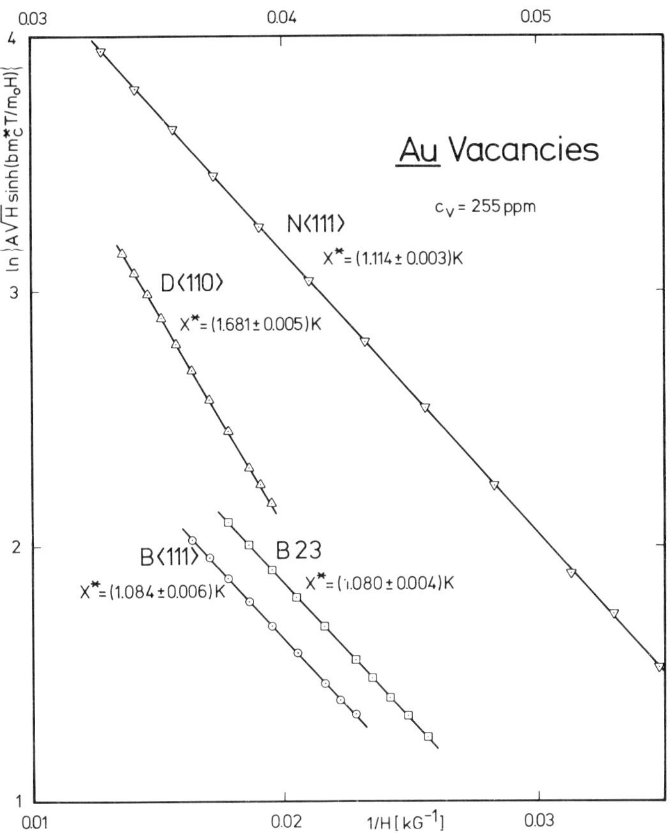

Fig. 6.1. Dingle plots for four extremal cross sections in gold containing 255 ppm of lattice vacancies. The slope of the lines is $-bm_c^*X^*/m_0$. The lower abscissa holds for the bellies and dogsbone and the upper for the neck orbit

centration. In our CuH and Au vacancy measurements /6.8,11/, we have measured the Dingle temperatures for 11 different hydrogen and vacancy concentrations. Fig. 6.2 shows some Dingle temperatures for vacancies in gold as a function of the vacancy concentration. The vacancies present in thermal equilibrium at high temperatures (600 - 1000° C) are quenched into the samples by quick cooling /6.13/. In the quenching process, dislocations are created which give rise to additional scattering and which manifest themselves as an intercept in Fig. 6.2 at $c_v = 0$. Dilute alloys can show such intercepts as well, as shown by BROWN et al. in AgAu and AgSn /6.9/. Here also the intercept is mainly due to dislocations created during the crystal pulling process. Since in many investigations the Dingle temperatures have been measured only for one concentration of scattering centers, the Dingle temperatures per at % are rather uncertain. Table 6.1 summarizes the Dingle temperatures for typical or-

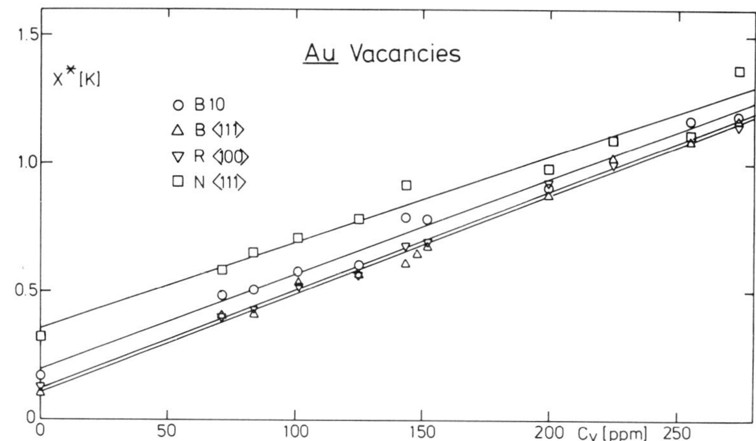

Fig. 6.2. Concentration dependence of the Dingle temperatures in the system Au vacancy. The linearly increasing part of X^* is due to the scattering of the conduction electrons by the vacancies and the lattice distortion surrounding them. The intercept at $c_V = 0$ is due to the scattering at dislocations which are created in the Au crystal during the quenching process

Table 6.1. Dingle temperatures in K/at% for some nonmagnetic scattering centers in Cu, Ag, and Au. The Dingle temperatures are quoted only for the stationary cross sections in the plane {110}. The residual resistivity ρ, the chemical valence difference ΔZ, and the charge ΔZ_B to be screened are quoted as well

X^* K/at%	CuAu /6.6/	CuNi /6.6/	CuGe /6.6/	CuH /6.8/	AgAu /6.9/	AgGe /6.9/	AgSn /6.9/	AuAg /6.1/	AuCu /6.1/	AuZn /6.1/	Au vacancy /6.11/
B<100>	13.2	26.6	109	37.9	8.7	300	212	9.8	9.2	35.6	38.3
BTP	13.9	28.8	119	48.4	8.7	273	210				38.0
B<111>	15.1	30.9	114	65	8.7	224	193	9.3	10.1	38.7	38.7
D<110>	12.4	26.6	161	77.3	7.0	332	272	7.6	8.6	36.3	37.0
R<100>	10.2	25.0	119	89.8	7.0	203	212	9.0	9.8	39.8	38.6
N<111>	7.9	14.9	188	112.6	3.6	351	318	2.8	4.5	24.5	33.8
ρ [$\mu\Omega$cm/at%]	0.55	1.11	3.79	1.50	0.38	5.5	4.3	0.36	0.45	0.95	1.69
ΔZ	0	-1	3	1	0	3	3	0	0	1	-1
ΔZ_B	-0.32	-0.94	2.81	0.89	0	3	2.76	0	0.21	1.1	-0.6

bits in a few Cu, Ag, and Au systems. All the defects are substitutional except for hydrogen which occupies octahedral interstices /6.8/. The different defect sites in an f.c.c. lattice are shown in Fig. 6.3. The scattering strength is nearly pro-

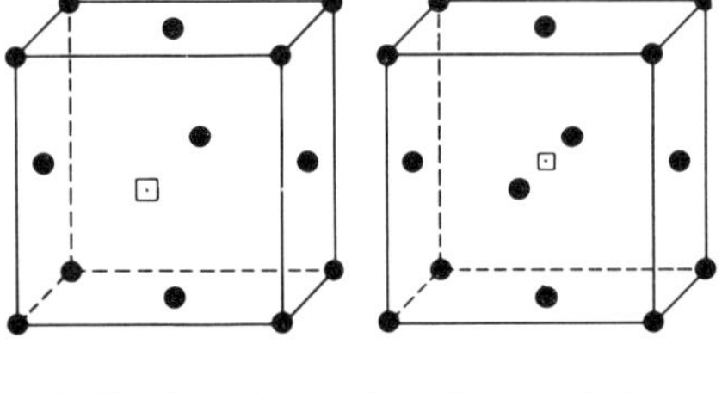

Fig. 6.3. Defects on a lattice site and on an octahedral interstice in an f.c.c. lattice

portional to the residual resistance ρ of the defects. On the other hand, the anisotropy of the scattering depends essentially on the site occupied by the defects. Substitutional defects with a small valence difference compared to the host lattice, $|\Delta Z| = 0.1$, scatter more strongly the belly electrons than the neck electrons. Hydrogen occupying octahedral interstices scatters the neck electrons more strongly. Local values of the scattering rates $1/\tau^*(\underline{k})$ can be obtained by deconvolution of Dingle temperatures according to (6.5). A symmetrized Fourier series expansion with fitting coefficients $T_{\ell mn}$ has been used to parameterize the scattering rates

$$\begin{aligned} 1/\tau^*(k) = {} & T_{000} + T_{110} \sum \cos \frac{a}{2}k_x \cos \frac{a}{2}k_y \\ & + T_{200} \sum \cos ak_x + T_{211} \sum \cos ak_x \cos \frac{a}{2}k_y \cos \frac{a}{2}k_z + \dots \quad . \end{aligned} \tag{6.15}$$

Local scattering rates for three characteristic systems from Table 6.1 are shown in Fig. 6.4 for some high symmetry directions. Ag in Au /6.1/ is a weak scattering center. The chemical valence difference is zero and the lattice distortion produced by the defect is negligible. Ag in Au is therefore a short-range scattering center on a substitutional lattice site. It scatters preferentially the belly electrons. The absolute values of the scattering rates are rather small. Hydrogen in copper /6.8/ constitutes a short-range potential on an octahedral interstice. It scatters preferentially the neck electrons. The overall scattering rates are larger than for a substitutional defect with $|\Delta Z| = 1$ (like Zn in Au or Ni in Cu). Finally, vacancies in gold are characterized by an especially strong distortion field. The atoms surrounding the vacancy relax into the empty lattice site so that the effective volume of the vacancy is not 1 but only 0.6 atomic volumes /6.11,14/. For vacancies

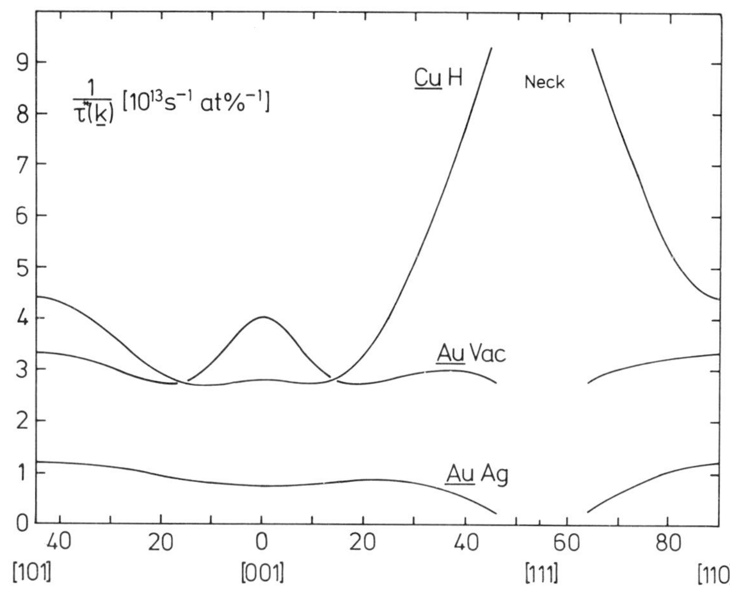

Fig. 6.4. Local scattering rates of the conduction electrons in AuAg, CuH, and Au vacancy. Ag is a well-localized defect on a lattice site in Au and scatters the neck electrons only weakly. H is a well-localized defect on an octahedral interstice in Cu and scatters most the neck electrons. Vacancies in Au create a strong distortion around the defect which causes the scattering anisotropy to be only weak

in gold, the scattering anisotropy is only weak. AuAg and CuH are two extreme cases of pronounced scattering anisotropy. The anisotropy of all other systems in the Table 6.1 is between these two extremes. Obviously the position and the strength of the defect influence the absolute values and the anisotropy of the scattering. These relations can be described quantitatively by a phase shift analysis of the Bloch waves of the host lattice at the defects.

6.3 Phase Shift Analysis of the Scattering of Conduction Electrons at Defects in the Noble Metals

MORGAN has given a formulation of the scattering of Bloch waves at a lattice defect /6.15/. In a series of papers, COLERIDGE, HOLZWARTH, and LEE have extended this theory to the description of scattering rates and Dingle temperatures in dilute noble metal alloys /6.16-22/. The results of Section 6.2 can be explained quantitatively within the framework of this theory. In the following, it will be assumed that the host lattice and the defects can be treated in the muffin-tin approximation (spherically symmetric potential inside a muffin-tin sphere of radius R_S and constant potential outside).

At first, the scattering of a plane wave at a single muffin-tin potential in vacuum will betreated. For a spherically symmetric potential, the solution of the Schrödinger equation can be separated into partial waves and each partial wave factored into a radial and spherical harmonic contribution. The asymptotic solution of the Schrödinger equation is written

$$\sum_L i^{\ell} a_L(\underline{k})[j_{\ell}(\kappa r) + i \sin \eta_{\ell} \exp(i\eta_{\ell}) h_{\ell}(\kappa r)]\ Y_L(\hat{\underline{r}}) \tag{6.16}$$

where $L \equiv \ell, m$, $E = \hbar^2\kappa^2/2m_0$ is the energy of the wave measured from the muffin-tin zero, j_{ℓ} and h_{ℓ} are spherical Bessel and Hankel functions, and $Y_L(\hat{\underline{r}})$ are spherical harmonics. Due to the conservation of angular momentum in a scattering process at a spherically symmetric potential, the waves with different ℓ do not mix. The scattering potential $V(r)$ scatters the incoming wave $j_{\ell}Y_L$. An outgoing spherical wave $h_{\ell}Y_L$ originates from the scattering center with amplitude $i \sin \eta_{\ell} \exp(i\eta_{\ell})$. The phase shifts η_{ℓ} depend on the potential through an element of the t-matrix

$$\begin{aligned} t_{\ell} &\equiv \sin \eta_{\ell} \exp(i\eta_{\ell}) \\ &= -(2m_0\kappa/\hbar^2) \int_0^{R_S} j_{\ell}(\kappa r) V(r) R^i_{\ell}(\kappa, r) r^2\, dr \end{aligned} \tag{6.17}$$

where $R^i_{\ell}(\kappa, r)$ is the solution of the radial Schrödinger equation inside the muffin-tin sphere.

In an ideal crystal lattice, the potentials of the lattice atoms are again approximated by muffin-tin potentials. In close-packed lattices with high electron concentrations like the noble metals, this assumption is rather well justified. The amplitudes a_L and b_L of the wave $a_L j_{\ell}(\kappa|\underline{r} - \underline{R}_N|)$ entering the N-th cell at $\underline{R}_N$ and of the wave $b_L h_{\ell}(\kappa|\underline{r} - \underline{R}_N|)$ leaving it are related again by

$$b_L/a_L = i \sin \eta^h_{\ell} \exp(i\eta^h_{\ell}) = it^h_{\ell} \tag{6.18}$$

where ℓ and m are no longer good quantum numbers. The nonspherical symmetry of the crystal potential removes partially the m degeneracy. In a cubic crystal, the wave functions are classified according to the angular momenta ℓ and the irreducible representations Γ of the cubic point group O_h belonging to the different ℓ. The first four are Γ_1 for $\ell = 0$, Γ_{15} for $\ell = 1$ and Γ_{12} and $\Gamma_{25'}$ for $\ell = 2$. In the following, $L \equiv \ell, \Gamma$ denotes this new set of quantum numbers. Whereas for the scattering at an isolated potential the amplitude a_L is a free parameter, it is built up in the lattice by superposition of the scattered waves which emerge from all other lattice sites.

$$a_L = i \sum_{L'} B_{LL'} b_{L'} \ . \tag{6.19}$$

The structure factors $B_{LL'}$ depend on $\underline{k}$, κ, and on the lattice but not on the phase shifts η_ℓ^h. It should be noted that waves with different L mix because the lattice is not spherically symmetric as seen from a muffin-tin cell. Combining (6.18) and (6.19) gives the KKR-band structure equations

$$\sum_{L'} (B_{LL'}(\underline{k},\kappa) t_{\ell'}^h + \delta_{LL'}) a_{L'} = 0 \ . \tag{6.20}$$

Nontrivial solutions exist only if

$$\det |B_{LL'} t_{\ell'}^h + \delta_{LL'}| = 0 \ . \tag{6.21}$$

The structure matrices $B_{LL'}$ are known for the noble metals. The Fermi energy E_F and the phase shifts $\eta_\ell^h(E_F)$ are determined in such a way that the radii $\underline{k}(E_F)$ fit optimally the dHvA measurements. In this way, the set of parameters (E_F, η_ℓ^h) in Table 4.2 has been determined. The amplitudes $a_L(\underline{k})$ of the L-th partial wave can now be calculated from (6.20). $|a_L(\underline{k})|^2$ describes the anisotropy of the wave character of the conduction electrons in a lattice cell.

In a third step a defect is imbedded in an otherwise ideal lattice. The defect potential is also described in the muffin-tin approximation. In the following substitutional and interstitial defects are treated separately.

6.3.1 Substitutional Defect

Consider a situation where a defect replaces a lattice atom in the N'-th cell. According to MORGAN /6.15/, the wave $j_\ell(\kappa|\underline{r} - \underline{R}_{N'}|)$ coming into this cell no longer has the amplitude a_L. Because the wave $b_L h_\ell(\kappa|\underline{r} - \underline{R}_{N'}|)$ emerging from N' is no longer a Bloch state, it creates by scattering at the surrounding lattice an additional contribution to the incoming wave. This effect is called backscattering and constitutes the essential difference between the scattering of plane waves at a defect in vacuum and the scattering of Bloch waves at a defect in a lattice. If the crystal can be treated nonrelativistically and if only phase shifts with $\ell \leqslant 2$ have to be considered, the backscattering can be taken into account by renormalizing the amplitude a_L by a complex factor A_L. There is a close relationship between A_L and the structure matrices $B_{LL'}$. Both describe how the emerging spherical waves superpose to construct an incoming wave, $B_{LL'}$ in the ideal lattice and A_L for the defect cell. A_L depends on the defect and on the host lattice

$$A_L(E_F) = \sin^2 \eta_\ell^h \exp(-i\Delta\eta_\ell) (\sin \eta_\ell^h \sin \eta_\ell^i - \chi_L \sin \Delta\eta_\ell)^{-1} \ . \tag{6.22}$$

Here

$$\Delta\eta_\ell = \eta_\ell^i - \eta_\ell^h \tag{6.23}$$

is the difference of the phase shifts for the defect and for the host atoms at the Fermi energy E_F. The Brillouin zone integral

$$\chi_L = (2\pi)^{-3} V_{EZ} \int_{BZ} d^3\underline{k} \, (B_{LL} + 1/t_\ell^h)^{-1} \tag{6.24}$$

expresses the close relationship between $B_{LL'}$ and A_L. In the free electron gas, where there is no backscattering ($B_{LL'} = 0$, $A_L = 1$), this integral is

$$\chi_L^{FE} = t_\ell^h = \sin \eta_\ell^h \exp(i\eta_\ell^h) \ . \tag{6.25}$$

The complex integrals χ_L depend only on the properties of the host lattice and the position of the defect, and not on the phase shifts η_ℓ^i of the defect. The values of χ_L have been calculated for Cu, Ag, and Au by LEE, HOLZWARTH and COLERIDGE /6.21/. An equivalent representation of A_L is

$$\begin{aligned} A_L(E_F) &= \sin^2 \eta_\ell^h \exp(-i\Delta\eta_\ell)[(\xi_\ell - \chi_L) \sin \Delta\eta_\ell]^{-1} \\ &= |A_L| \exp(i\Theta_L) \end{aligned} \tag{6.26}$$

with

$$|A_L(E_F)| = \sin^2 \eta_\ell^h (\sin \Delta\eta_\ell)^{-1} [(\xi_\ell - \mathrm{Re}\{\chi_L\})^2 + (\mathrm{Im}\{\chi_L\})^2]^{-1/2} \tag{6.27}$$

$$\Theta_L = \mathrm{arctg}\,[\mathrm{Im}\{\chi_L\}/(\xi_\ell - \mathrm{Re}\{\chi_L\})] - \Delta\eta_\ell \tag{6.28}$$

$$\xi_\ell = (\mathrm{cotg}\, \eta_\ell^h - \mathrm{cotg}\, \eta_\ell^i)^{-1} \ . \tag{6.29}$$

The factor A_L renormalizes the amplitude and the phase of the L-th wave entering the defect cell. Thus the total phase shift Φ_L produced by the defect contains two contributions. There is first the usual phase shift $\Delta\eta_\ell$ between the outgoing and the incoming waves. In addition, the incoming wave is already phase shifted by an amount Θ_L compared to a wave entering an ideal lattice cell. The total phase shift Φ_L is called Friedel phase shift

$$\begin{aligned} \Phi_L &= \Delta\eta_\ell + \Theta_L \\ &= \mathrm{arctg}[\mathrm{Im}\{\chi_L\}/(\xi_\ell - \mathrm{Re}\{\chi_L\})] \ . \end{aligned} \tag{6.30}$$

The conduction electrons create by scattering a charge transfer $\widetilde{F}$ which screens the defect-induced charge ΔZ_B and satisfies the Friedel sum rule

$$\widetilde{F} \equiv \sum_L 2n_L \Phi_L/\pi = \Delta Z_B \tag{6.31}$$

where n_L is the degeneracy of the representation L. [$n(0,\Gamma_1) = 1$, $n(1,\Gamma_{15}) = 3$,

$n(2,\Gamma_{12}) = 2$, and $n(2,\Gamma_{25'}) = 3$].

The scattering rate $1/\tau(\underline{k})$ is related to the diagonal elements $T_{\underline{k}\underline{k}}$ of the t-matrix by (6.13). The analogue expression to (6.17) is /6.15/

$$T_{\underline{k}\underline{k}}(E_F) = -(\hbar^2/2m_o\kappa) \sum_L |a_L(\underline{k},E_F)|^2 A_L(E_F) \sin \Delta\eta_\ell \exp(i\Delta\eta_\ell) . \tag{6.32}$$

Each partial wave contribution in (6.32) is the product of an anisotropy factor t_k^L and a scattering parameter S_L. t_k^L depends only on the host lattice and can be calculated from the solution of the KKR equation (6.20),

$$t_k^L = (\hbar^2/2m_o\kappa) \, |a_L(\underline{k})|^2 \sin^2 \eta_\ell^h . \tag{6.33}$$

The scattering parameter S_L

$$S_L = A_L \sin \Delta\eta_\ell \exp(i\Delta\eta_\ell)(\sin \eta_\ell^h)^{-2} = (\xi_\ell - \chi_L)^{-1} \tag{6.34}$$

is $\underline{k}$-independent, describes the properties of the defect, and depends on the host lattice through η_ℓ^h and χ_L. According to (6.13) the scattering rate is

$$1/\tau(\underline{k}) = (2c_d/\hbar) \sum_L t_k^L \, \mathrm{Im}\{S_L\} \tag{6.35}$$

$$\mathrm{Im}\{S_L\} = \mathrm{Im}\{\chi_L\} \, [(\xi_\ell - \mathrm{Re}\{\chi_L\})^2 + (\mathrm{Im}\{\chi_L\})^2]^{-1} . \tag{6.36}$$

Equations (6.33,35,36) are the essential result of the phase shift analysis for the scattering of Bloch waves at a defect. The scattering anisotropy of the individual partial waves L depends on the host lattice and is given by the factor t_k^L. $\mathrm{Im}\{S_L\}$ is the weight factor with which the rates of the different L are superposed to form the total scattering rate. It depends on the defect (η_ℓ^i) and on the host lattice (η_ℓ^h,χ_L). Using (6.4) and (6.35), the Dingle temperatures can be written

$$Xm_c/m_o = (\varepsilon_o/k_B)(a_o/a)^2 c_d \sum_L W_L \, \mathrm{Im}\{S_L\} . \tag{6.37}$$

Here ε_o is 1 Rydberg, a_o is the Bohr radius, and a the lattice parameter.

$$W_L = (a/\pi)^2 \oint dk \; t_k^L/\hbar v_\perp \tag{6.38}$$

is the average of the anisotropy factors t_k^L over the appropriate extremal cross section. The coefficients W_L have been calculated for stationary extremal cross sections and for substitutional defects in Cu, Ag, and Au /6.17,20/. They are listed in Table 6.2 for some orbits in Cu. It can be seen that the neck electrons have mainly p-character at a lattice site with no s-character and a little d-admixture. The B<100> electrons have mainly p- and d-character.

In the present description, the lattice and the defect have been treated nonrelativistically. For elements with a high Z as host (e.g., Au) or as defect, the spin-

Table 6.2. Brillouin zone integrals χ_L, $\tilde{\chi}_L$ and averages W_L, $\tilde{W}_L$ of the anisotropy factors t_k^L, $\tilde{t}_k^L$ for some extremal cross sections in Cu for substitutional defects (S) and for defects on octahedral interstices (O). t_k^L and $\tilde{t}_k^L$ give the contribution of the wave character L ($s\Gamma_1$, $p\Gamma_{15}$, $d\Gamma_{12}$, $d\Gamma_{25'}$) in the Bloch wave with wave vector $\underline{k}$ at the position of the defect

	site	$0\Gamma_1$	$1\Gamma_{15}$	$2\Gamma_{12}$	$2\Gamma_{25'}$
$\mathrm{Re}\{\chi_L\}$	S	0.07607	0.12412	-0.13233	-0.12862
$\mathrm{Re}\{\tilde{\chi}_L\}$	O	-0.51768	-0.92621	-5.55897	-5.61006
$\mathrm{Im}\{\chi_L\}$	S	0.00318	0.01392	0.01502	0.01564
$\mathrm{Im}\{\tilde{\chi}_L\}$	O	0.74896	0.86023	0.45958	1.36575
W_L, $\tilde{W}_L$					
B<100>	S	0.01317	0.12156	0.09939	0.14539
	O	1.42530	12.59223	5.00423	3.61711
B<111>	S	0.01144	0.10998	0.10695	0.18667
	O	2.39290	6.63995	1.90415	11.51961
D<110>	S	0.00817	0.14234	0.08301	0.11562
	O	2.45904	7.84367	2.47674	14.68939
R<100>	S	0.00799	0.13371	0.09156	0.14747
	O	2.98703	4.80336	1.17912	17.43271
N<111>	S	0	0.07396	0.02020	0.01872
	O	1.59115	0.66813	0.09496	12.10754

-orbit coupling must be taken into account. The relativistic formulation of the phase shift analysis has been given by HOLZWARTH and LEE /6.20/. Since it is similar to the nonrelativistic one, it will not be discussed further here.

The host lattice parameters η_ℓ^h, χ_L, and W_L have been determined for the noble metals from nonrelativistic and from relativistic band structure calculations. Experimental Dingle temperatures can be parametrized according to (6.36-38). By means of nonlinear least squares fitting procedures (for instance the program VA05AD from the Harwell Subroutine Library), the defect phase shifts η_ℓ^i are chosen in such a way that the calculated Dingle temperatures agree with the measured ones. The Friedel phases Φ_L, the backscattering phase shifts Θ_L, and the Friedel sum $\tilde{F}$ can be

Table 6.3. Phase shift analysis for three typical dilute alloys in Table 6.1. Ni in Cu scatters most strongly the d-waves, Ag in Au the s-wave, and Ge in Cu the p-wave. Backscattering creates a substantial contribution in the Friedel phases. The Friedel sum $\tilde{F}$ and the charge ΔZ_B agree quite well for CuNi and for AuAg; not so for CuGe where the lattice distortion is rather high

	ΔZ	L	Φ_L	η^i_ℓ	$\Delta\eta_\ell$	Θ_L	$\lvert A_L\rvert$	$\tilde{F}$	ΔZ_B
CuNi	-1	$0\Gamma_1$	0.00	0.076	0	0	1	-0.96	-0.94
		$1\Gamma_{15}$	-0.061	0.058	-0.072	0.01	1.02		
		$2\Gamma_{12}$	-0.26	-0.31	-0.19	-0.07	1.24		
		$2\Gamma_{25'}$	-0.27	-0.32	-0.20	-0.07	1.18		
AuAg	0	$0\Gamma_1$	-0.26	-0.28	-0.53	0.27	0.75	-0.13	0
		$1\Gamma_{15}$	-0.067	-0.06	-0.12	0.06	1.03		
		$2\Gamma_{12}$	0.04	-0.20	0.048	-0.01	0.94		
		$2\Gamma_{25'}$	0.06	-0.19	0.05	0.01	0.94		
CuGe	3	$0\Gamma_1$	0.09	0.23	0.16	-0.07	1.03	2.14	2.81
		$1\Gamma_{15}$	0.73	1.11	0.98	-0.25	0.97		
		$2\Gamma_{12}$	0.21	0.13	0.25	-0.04	0.80		
		$2\Gamma_{25'}$	0.22	0.12	0.24	-0.02	0.84		

calculated from (6.29-31). The experimental data of Table 6.1 have been analyzed in this way. Table 6.3 shows the results of fits for three typical substitutional alloys (CuNi , AuAg, and CuGe). The first column ΔZ gives the chemical valence difference between the impurity and the host lattice. The Friedel phase shifts indicate that Ni in Cu is mainly a d-scatterer, Ag in Au is mainly an s-scatterer, and Ge in Cu is a particularly strong p-scatterer. Backscattering creates significant differences between the Φ_L and $\Delta\eta_\ell$. In the system AuAg, for instance, the s-wave is shifted by backscattering by 0.27 rad and reduced in amplitude by 25%. The charge transfer due to the scattering is given by the Friedel sum $\tilde{F}$. It must screen the defect-induced charge ΔZ_B. If the defect has no lattice distortion, ΔZ_B is equal to ΔZ. But in general the defect distorts the lattice. ESHELBY /6.23/ has shown that a center of dilatation in a finite anisotropic continuum with a stress-free surface creates a lattice dilatation ΔV_{fin} which contains a contribution ΔV_{loc} confined to the impurity site and a long-range contribution from the relaxation of the stress-free surface.

The relative volume change $(\Delta V/V)_{fin}$ is given by the relative lattice parameter change $3\Delta a/a$ and the local volume change is given in terms of $\Delta a/a$ by

$$(\Delta V/V)_{loc} = 3\Delta a/\tilde{\gamma}a \ . \qquad (6.39)$$

The constant $\tilde{\gamma}$ has the values 1.45 , 1.42 , and 1.23 for Cu, Ag, and Au, respectively /6.24/. BLATT has shown that the charge to be screened by the conduction electrons is approximately /6.25/

$$\Delta Z_B = \Delta Z - (\Delta V/V)_{loc} \ . \qquad (6.40)$$

When the lattice distortion is not too large, the Friedel sum agrees quite well with ΔZ_B. For strong lattice distortion, the phase shift analysis in the present form is no longer adequate for the description of the scattering. Up to now, it has been assumed that the defect is confined to a muffin-tin sphere in an ideal lattice. If this is no longer appropriate for reasons of strong lattice distortion, the defect potential and the backscattering are changed. These effects are not included in the existing theory. An example of a defect with strong lattice distortion is the vacancy in Au /6.11/. The local volume change $(\Delta V/V)_{loc} = -0.40$, i.e., by lattice relaxation the volume of a vacancy is reduced to 0.60 atomic volume. The phase shift analysis no longer works for this system. Model calculations for a vacancy in Au have shown that the neck scattering rates for a "vacancy" without distortion are only about half as large as the belly scattering rates /6.11/. The lattice relaxation particularly enhances the neck scattering rates so that the measured scattering rates show only very little anisotropy (Fig. 6.4). A possibility to separate the effects of the vacancy and of the surrounding distortion field on the scattering rates is given in Section 6.4.

6.3.2 Defects on Octahedral Interstices

A defect influences the scattering rates not only by the phase shifts η_ℓ^i but also by its position in the lattice. The contribution of the different partial waves L to the scattering rates $1/\tau(\underline{k})$ (6.35) is the product of two factors. t_k^L gives the contribution of the L-th wave in the Bloch wave vector $\underline{k}$ at the position of the defect. Since the Bloch wave at a lattice site is different from that at an interstitial position, t_k^L depends on the position of the defect. The same holds for the scattering parameter S_L. The superposition of the scattered waves which are reflected from the surroundings has to be considered at the defect position. Hence, the Brillouin zone integrals χ_L and S_L depend on the defect position. HOLZWARTH and LEE /6.22/ have formulated the phase shift analysis for the case of a defect occupying octahedral interstices in Cu. The relations for the renormalization factors $\tilde{A}_L$, for the scattering rates $1/\tau(\underline{k})$, and for the Dingle temperatures X are now

$$\tilde{A}_L = [\sin \eta_\ell^i \exp(i\eta_\ell^i)(\tilde{\xi}_\ell - \tilde{\chi}_L)]^{-1} = |\tilde{A}_L| \exp(i\tilde{\Theta}_L) \tag{6.41}$$

with

$$|\tilde{A}_L| = [\sin \eta_\ell^i \{(\tilde{\xi}_\ell - \mathrm{Re}\{\tilde{\chi}_L\})^2 + (\mathrm{Im}\{\tilde{\chi}_L\})^2\}^{1/2}]^{-1} \tag{6.42}$$

$$\tilde{\Theta}_L = \mathrm{arctg}[\mathrm{Im}\{\tilde{\chi}_L\}/(\tilde{\xi}_\ell - \mathrm{Re}\{\tilde{\chi}_L\})] - \eta_\ell^i = \tilde{\Phi}_L - \eta_\ell^i \tag{6.43}$$

$$1/\tau(\underline{k}) = (2c_d/\hbar)\sum_L \tilde{t}_{\underline{k}}^L \,\mathrm{Im}\{\tilde{S}_L\} \tag{6.44}$$

$$Xm_c/m_o = (\varepsilon_o/k_B)(a_o/a)^2 c_d \sum_L \tilde{W}_L \,\mathrm{Im}\{\tilde{S}_L\} \tag{6.45}$$

$$\tilde{\xi}_\ell = \mathrm{cotg}\, \eta_\ell^i \,. \tag{6.46}$$

The quantities marked by a tilde depend on the position of the defect in the lattice. The coefficients $\tilde{W}_L$ and $\tilde{\chi}_L$ calculated by HOLZWARTH and LEE for octahedral interstices in Cu are listed in Table 6.2. It should be emphasized that W_L and $\tilde{W}_L$ have different magnitudes and different anisotropy. The neck electrons, for instance, have mainly p-character on substitutional sites and mainly s- and $d\Gamma_{25'}$-character on octahedral interstices. The analysis of the CuH results is given in Table 6.4. The hydrogen on

Table 6.4. Phase shift analysis for hydrogen in Cu occupying octahedral interstices

	η_ℓ^i	$\tilde{\Phi}_L$	$\tilde{\Theta}_L$	$\lvert\tilde{A}_L\rvert$
$0\Gamma_1$	1.641	1.032	-0.609	1.149
$1\Gamma_{15}$	0.152	0.115	-0.037	0.880
$2\Gamma_{22}$	0	0	0	1
$2\Gamma_{25'}$	0	0	0	1

an octahedral interstice is a defect of short range creating mainly s-scattering. That is the reason for the stronger neck and the weaker B<100> scattering (Fig. 6.4). The Friedel phases create a charge transfer of $\tilde{F}$ = 0.88, in good agreement with the charge to be screened ΔZ_B = 0.89 /6.8/. The backscattering changes the amplitudes of the incoming s- and p-waves by +15% and -12%. The phase shifts $\tilde{\Theta}_L$ are appreciable. The phase shifts η_ℓ^i alone would give a Friedel sum of 1.34. This differs from the real Friedel sum by 0.46 and corresponds to the charge transfer by backscattering.

6.3.3 Scattering of the Conduction Electrons by Hydrogen in Cu Occupying Octahedral Interstices and Lattice Sites

The influence of the position of the defects on the Dingle temperatures can be illustrated by hydrogen occupying octahedral interstices and lattice sites in Cu. Loading copper with hydrogen at temperatures around $650^{\circ}C$ leads to the occupation of octahedral interstices /6.26/. But there are indications that loading near the melting point of copper can lead to the occupation of lattice sites by the hydrogen. It is known that hydrogen can be trapped by impurities at low temperatures ($-100^{\circ}C$) /6.26/. On the other hand, it is possible to quench hydrogen and vacancies into copper by appropriate loading conditions /6.27/. If the hydrogen could be trapped by the vacancies (thereby becoming a substitutional defect), this would have substantial consequences for the scattering of the conduction electrons. Using the phase shifts η_{ℓ}^{i} of Table 6.4 for both lattice sites, the Dingle temperatures listed in Table 6.5 are obtained. They differ appreciably for the two positions especially

Table 6.5. Dingle temperatures in CuH for hydrogen occupying octahedral interstices and lattice sites, calculated with the phase shifts of Table 6.4. The neck scattering rates are especially strongly affected by the position of the defect

X^* K/at.%	Octahedral interstice	Lattice site
B<100>	40.3	56.8
BTP	48.4	54.3
B<111>	61.2	49.9
D<110>	68.7	38.7
R<100>	78.5	37.6
N<111>	120.7	2.8

for the neck orbits. It is obvious that measurements of the Dingle temperatures give the possibility to determine whether hydrogen can be trapped by a vacancy.

6.4 Phase Shift Analysis of Defect-Induced Fermi Surface Changes

Besides reducing the lifetime of conduction electrons, defects also alter the geometry of the Fermi surface /2.10,6.19,28/. When each impurity is screened separately by the conduction electrons (dilute limit) and when the defects create no lattice distortion, the Fermi energy is not changed by the defects. Indeed, in this case the lattice is undisturbed between the defects and, since the Fermi energy must be constant throughout the lattice, its value remains that of the ideal lattice. On the other hand, the geometry of the Fermi surface is changed by the defects. If $\underline{k}$ is the wave vector in the perfect crystal, the wave vector $\underline{q}$ in the defect lattice is given by /6.19,29/

$$\underline{q} = \underline{k} - c_d \, d\underline{k}/dE \, \mathrm{Re}\{T_{\underline{k}\underline{k}}\} \tag{6.47}$$

or using (6.32-34)

$$\underline{q} = \underline{k} + c_d \, d\underline{k}/dE \sum_L t_k^L \, \mathrm{Re}\{S_L\} \tag{6.48}$$

where

$$\mathrm{Re}\{S_L\} = (\xi_\ell - \mathrm{Re}\{\chi_L\}) \, [(\xi_\ell - \mathrm{Re}\{\chi_L\})^2 + (\mathrm{Im}\{\chi_L\})^2]^{-1} \; . \tag{6.49}$$

The similarity of (6.48,49) with (6.35,36) for the scattering rates is obvious. The real part of T_{kk} is correlated with the defect-induced Fermi surface changes and the imaginary part with the lifetime of the electrons. The defect-induced changes of the dHvA frequencies are obtained from (6.48) by averaging over an extremal cross section

$$\Delta F = (\hbar c/2\pi e_0)(\pi/a)^2 \, c_d \sum_L W_L \, \mathrm{Re}\{S_L\} \; . \tag{6.50}$$

The coefficients W_L are given by (6.38). Thus the Dingle temperature X and the Fermi surface changes ΔF can both be described by the phase shifts η_ℓ^i and Φ_L.

Two differences in the determination of η_ℓ^i and Φ_L from X and ΔF should be emphasized. First, the phase shifts η_ℓ^i can be determined from ΔF including their sign, in contrast to those determined from Dingle temperatures. The reason is that $\mathrm{Re}\{S_L\}$ contains the ξ_ℓ linearly while $\mathrm{Im}\{S_L\}$ contains them quadratically. In the analysis of the Dingle temperatures, the sign of η_ℓ^i must be determined from other criteria, such as the Friedel sum rule or the resistivity.

The second difference in the information contained in the Dingle temperatures and in the Fermi surface changes concerns the influence of the defect-induced lattice distortion. In the simplest approach, the lattice distortion created by defects is treated in an elastic continuum model. Within this model the lattice expansion ΔV_{fin} in a finite crystal with stress-free surface can be separated into a dilatation ΔV_{loc} confined to the defect site and a long-range expansion ΔV_I due to the relaxation of the stress-free surface /6.23,24/

$$\Delta V_{fin} = \Delta V_{loc} + \Delta V_I \; . \tag{6.51}$$

ΔV_{fin} is related to the lattice parameter change

$$(\Delta V/V)_{fin} = 3\ \Delta a/a\ . \qquad (6.52)$$

There exists experimental evidence /6.19/ that the local lattice distortion around the defects contributes to the dHvA frequency changes ΔF only through its contribution to the homogeneous lattice expansion expressed by the lattice parameter change. Since $\Delta a/a$ is known experimentally for many alloys the defect-induced lattice distortion can be taken into account rather easily in the determination of phase shifts from the frequency changes ΔF. In contrast to this, the local lattice distortion affects the scattering rates in a more subtle way, by changing the scattering potential and the backscattering from the atoms surrounding the defect. It has not yet been possible to treat these changes in a satisfactory way. Thus the Friedel phase shifts derived from Dingle temperatures differ in general from those derived from Fermi surface changes /6.19/. If this concept is confirmed by further experiments, a comparison of the phase shifts may help to clarify the influence of lattice distortions on the scattering rates.

Acknowledgements

I would like to thank the following persons who have contributed to this article:
Dr. W.R. Wampler, Prof. R.R. Bourassa, Dr. P.H. Dederichs and Dr. R.O. Jones for many helpful discussions,
Dr. W. Uelhoff, M. Abdel-Fattah and G. Hanke for the preparation of the high quality single crystals used in the experiments,
Dr. K. Mika and K. Wingerath for help in the numerical evaluation of the data.
Dr. J. Mundy, Dr. J.B. Roberto and Dr. E. Seitz for carefully reading the manuscript and Mrs. G. Hahn and Mrs. M. Klein for typing it.
Finally I would like to express my special gratitude to Prof. W. Schilling who has initiated and continuously supported our de Haas-van Alphen work.

List of Symbols

$\underline{A}$	vector potential
A, A_n	area of cyclotron orbit in $\underline{k}$-space
A_{ex}, $A^{\pm}_{ex}$	area of extremal cross section of the Fermi surface
$A(T,H,X^*)$	dHvA amplitude (2.43)
A_L, $\widetilde{A}_L$	backscattering renormalization factor (6.22,41)
a	lattice parameter
a_o	Bohr radius
$a_L(\underline{k})$	amplitude of the L-th partial wave $j_\ell Y_L$ (6.16)
$\underline{B}$	magnetic induction
$B_{LL'}$	structure matrix of KKR-band structure (6.19)
b	$2\pi^2 k_B c m_o/\hbar e_o$ = 146.925 kG/K (2.23)
$b_L(\underline{k})$	amplitude of the L-th partial wave $h_\ell Y_L$ (6.16)
$C_{\ell mn}$, $C^{\pm}_{\ell mn}$	coefficients of Fourier series representation of the Fermi surface (4.1)
c	velocity of light
c_d	concentration of defects
$D(E)$	density of states
$D(H)$	magnetization of electrons in extremal cross section (2.44)
d	degeneracy of Landau levels
E, E_n	electron energy
E_F	Fermi energy
e_o	charge of proton
F	dHvA frequency (2.16)
$\widetilde{F}$	Friedel sum (6.31)
G	oscillatory part of electron free energy
g_c	electron g-factor
$\underline{H}$, H_1, H_2	magnetic fields
h	modulation field amplitude
$\hbar$	Planck's constant
I_1	temperature damping factor of dHvA amplitudes (2.45)
K_1	Dingle damping factor (2.46)
$\underline{k}$, $\underline{k}_\perp$	wave vectors
k_F, k_s	radius of free electron Fermi surface
k_B	Boltzmann's constant
L	ℓm or ℓ,Γ
ℓ	angular momentum quantum number
M	oscillatory part of magnetization
m_o	free electron mass

m_c	cyclotron mass (2.7)
n	integer
n_{el}	electron density
$P(\underline{k},\underline{k}')$	scattering transition rate (6.10)
$\underline{q}$	wave vector in defect lattice
$\underline{r}$	radius vector
S_1	spin-splitting factor in dHvA amplitude (2.47)
S_L, $\tilde{S}_L$	scattering parameter (6.34)
S_n	area enclosed by cyclotron orbit in $\underline{r}$-space
s	spin quantum number
T	temperature
T_c	cyclotron period (2.5)
$T_{kk'}$	element of t-matrix (6.10)
t	time
t_ℓ, t_ℓ^h	t-matrix (6.17,18)
t_k^L	anisotropy factor of T_{kk} (6.33)
$\underline{v}$, $\underline{v}_\perp$	electron velocity
v_s	free electron Fermi velocity
W_L, $\tilde{W}_L$	orbital averages of t_k^L and $\tilde{t}_k^L$ (6.38)
X	Dingle temperature
α	angle characterizing a point on the cyclotron orbit
Γ	representations of the cubic group
γ^*, γ_s	coefficients of electronic specific heat (5.8)
ΔF	defect-induced changes of dHvA frequency
ΔH	dHvA period (2.18)
ΔV_{fin}, ΔV_{loc}, ΔV_I	defect-induced volume changes in a lattice
ΔZ	valence difference between defect and host
ΔZ_B	charge to be screened by conduction electrons
η_ℓ, η_ℓ^h, η_ℓ^i, $\Delta\eta_\ell$	scattering phase shifts
Θ	angle characterizing a cyclotron orbit in the {110} plane (Fig. 5.2)
Θ_D	Debye temperature
Θ_L, $\tilde{\Theta}_L$	backscattering phase shifts (6.28,43)
θ	polar angle
θ	scattering angle
κ	$(2m_oE/\hbar^2)^{1/2}$
$\hat{\kappa}$	curvature of extremal cross section along field direction
$\lambda(\underline{k})$	electron-phonon coupling constant (2.32)
$\bar{\lambda}$	electron-phonon coupling constant averaged over the Fermi surface

ν	integer
ξ_ℓ, $\tilde{\xi}_\ell$	(6.29) and (6.46)
ρ	residual resistivity produced by defects
$\tau(\underline{k})$	electron lifetime (6.11)
$\tau_{tr}(\underline{k})$	transport relaxation time (6.14)
Φ	angle characterizing a cyclotron orbit in the plane {100} (Fig. 5.2)
Φ_L, $\tilde{\Phi}_L$	Friedel phase shifts
ϕ_0	flux quantum
ϕ	polar angle
χ_L, $\tilde{\chi}_L$	Brillouin zone integrals (6.24,25)
ω	modulation frequency
ω_c	cyclotron frequency (2.4)
*	quantity renormalized by electron-phonon interaction
< >	orbital average of a quantity defined on the Fermi surface (2.34)

References

2.1 W.J. de Haas, P.M. van Alphen: Leiden Comm.208d, 212a (1930)
2.2 R. Peierls: Z. Phys. 81, 186 (1933)
2.3 L.D. Landau, appendix in D. Schoenberg: Proc. Roy. Soc. A170, 341 (1939)
2.4 L. Onsager: Phil. Mag. 43, 1006 (1952)
2.5 I.M. Lifshitz, A.M. Kosevich: J. Exp. Th. Phys. USSR 29, 730 (1955)
2.6 D. Shoenberg: "The de Haas-van Alphen Effect" in C.J. Gorter: Progress in Low Temperature Physics II, 226 (1957)
2.7 A.P. Cracknell: *The Fermi Surfaces of Metals* (London: Taylor & Francis 1971)
2.8 A.V. Gold: "The de Haas-van Alphen Effect", in J.F. Cochran, R.R. Haering: *Solid State Physics*, Vol. I, p. 38 (New York: Gordon & Breach 1968)
2.9 R.B. Dingle: Proc. Roy. Soc. (London) A 211, 517 (1952)
2.10 E. Mann: Phys. kondens. Materie 12, 210 (1971)
2.11 J.W. Wilkins: *Observable Many-Body Effects in Metals* (Copenhagen: Nordita 1968
2.12 J.M. Luttinger: Phys. Rev. 121, 1251 (1961)
2.13 S. Engelsberg, G. Simpson: Phys. Rev. B2, 1657 (1970)
2.14 F.M. Mueller, H.W. Myron: Commun. Phys. 1, 99 (1976)
3.1 D. Shoenberg: Phil. Trans. Roy. Soc. (London) A255, 85 (1962)
3.2 S.Foner: Rev. Sci. Instr. 30, 548 (1959)
3.3 A.S. Joseph, A.C. Thorsen: Phys. Rev. A133, 1546 (1964)
3.4 D. Shoenberg, P.J. Stiles: Proc. Roy. Soc. (London) A281, 62 (1964)
3.5 L.R. Windmiller, J.B. Ketterson: Rev. Sci. Instr. 39, 1672 (1968)
3.6 W.R. Wampler, S. Matula, B. Lengeler, G. Durcansky: Rev. Sci. Instr. 46, 58 (1975)
3.7 H. Fehmer, W. Uelhoff: J. Crystal Growth 13/14, 257 (1972)
3.8 D. Shoenberg: Canad. J. Phys. 46, 1915 (1968)
4.1 D. Shoenberg: Phil. Mag. 5, 105 (1960)
4.2 A.B. Pippard: Phil. Trans. Roy. Soc. (London) A250, 325 (1957)
4.3 D. Shoenberg: Phil. Trans. Roy. Soc. (London) A255, 85 (1962)
4.4 B. Bosacchi, J.B. Ketterson, L.R. Windmiller: Phys. Rev. B4, 1197 (1971)

4.5 P.T. Coleridge, I.M. Templeton: J. Phys. F: Metal Phys. 2, 643 (1972)
4.6 M.R. Halse: Phil. Trans. Roy. Soc. (London) A265, 507 (1969)
4.7 J.O. Dimmock: Solid State Physics 26, 103 (1971)
4.8 B. Segall, F.S. Ham: In *Method in Computational Physics, Energy Bands in Solids*, Vol. 8, Chap. 7, eds. A. Alder, S. Fernberg, M. Rotenberg (New York: Academic Press 1968)
4.9 G.A. Burdick, Phys. Rev. 129, 138 (1963)
4.10 B. Segall, Phys. Rev. 125, 109 (1962)
4.11 J.S. Faulkner, H.L. Davis, H.W. Joy: Phys. Rev. 161, 656 (1967)
4.12 E.C. Snow: Phys. Rev. 171, 785 (1968)
4.13 N.E. Christensen, B.O. Seraphin: Sol. State Commun. 8, 1221 (1970)
4.14 M.J.G. Lee, N.A.W. Holzwarth, P.T. Coleridge: Phys. Rev. B13, 3249 (1976)
5.1 B. Lengeler, W.R. Wampler, R.R. Bourassa, K. Mika, K. Wingerath, W. Uelhoff: Phys. Rev. B15, 5493 (1977)
5.2 The "1958 ^{4}He Scale of Temperature"; NBS Monograph No. 10
5.3 J.R. Klauder, W.A. Reed, G.F. Brennert, J.E. Kunzler: Phys. Rev. 141, 592 (1966)
5.4 J.F. Koch, R.A. Stradling, A.F. Kip: Phys. Rev. 133, A240 (1964)
5.5 D.G. Howard: Phys. Rev. 140, A1705 (1965)
5.6 D.N. Langenberg, S.M. Marcus: Phys. Rev. 136, A1383 (1964)
5.7 P.T. Coleridge, B.R. Watts: Canad. J. Phys. 49, 2379 (1971)
5.8 A.S. Joseph, A.C. Thorsen: Phys. Rev. 138, A1159 (1965)
5.9 R.E. Doezema, J.F. Koch: Phys. Rev. B5, 3866 (1972)
5.10 P. Deimel, R.E. Doezema: Phys. Rev. B10, 4897 (1974)
5.11 P.E. Lewis, P.M. Lee: Phys. Rev. 175, 795 (1968)
5.12 W.J. O'Sullivan, A.C. Switendick, J.E. Schirber: Phys. Rev. B1, 1443 (1970)
5.13 J.F. Janak, A.R. Williams, V.L. Moruzzi: Phys. Rev. B6, 4367 (1972)
5.14 M.G.J. Lee: Phys. Rev. B2, 250 (1970)
5.15 R.F. Hoyt, A.C. Mota: Solid State Comm. 18, 139 (1976)
5.16 P.M. Chaikin, P.K. Hansma: Phys. Rev. Lett. 36, 1552 (1976)
5.17 H. Teichler: Phys. stat. sol. (b) 48, 189 (1971)
5.18 S.G. Das: Phys. Rev. B7, 2238 (1973)
5.19 D. Nowak: Phys. Rev. B6, 3691 (1972)
5.20 G. Grimvall: Phys. kondens. Materie 14, 101 (1972)
5.21 D.L. Martin: Phys. Rev. B8, 5357 (1973)
5.22 B.Bosacchi, P. Franzosi: Phys. Rev. B12, 5999 (1975)
6.1 D.H. Lowndes, K.M. Miller, R.G. Poulsen, M. Springford: Proc. Roy. Soc. (London) A331, 497 (1973)
6.2 L.S. Rodberg, R.M. Thaler: *The Quantum Theory of Scattering*, p. 182 ff.(New York: Academic Press 1967)
6.3 H. Shiba: Phys. cond. Matter 19, 259 (1975)
6.4 P.T. Coleridge, B.R. Watts: Phil. Mag. 24. 1163 (1971)
6.5 Y.K. Chang, R.J. Higgins: Phys. Rev. B12. 4261 (1975)
6.6 R.G. Poulsen, D.L. Randles, M. Springford: J. Phys. F: Metal Phys. 4, 981 (1974)
6.7 I.M. Templeton, P.T. Coleridge: J. Phys. F: Metal Phys. 5, 1307 (1975)
6.8 W.R. Wampler, B. Lengeler: Phys. Rev. B15, 4614 (1977)
6.9 H.R. Brown, A. Myers: J. Phys. F: Metal Phys. 2, 683 (1972)
6.10 Y. Chung, D.H. Lowndes: J. Phys. F: Metal Phys. 6, 199 (1976)
6.11 B. Lengeler: Phys. Rev. B15, 5504 (1977)
6.12 Y.K. Chang, G.W. Crabtree, J.B. Ketterson: Phys. Rev. B16, 714 (1977)
6.13 B. Lengeler: Phil. Mag. 34, 259 (1976)
6.14 W. Hertz, W. Waidelich, H. Peisl: Phys. Lett. 43A, 289 (1973)
6.15 G.J. Morgan: Proc. Phys. Soc. 89, 365 (1966)
6.16 P.T. Coleridge: J. Phys. F: Metal Phys. 2, 1016 (1972)
6.17 P.T. Coleridge, N.A.W. Holzwarth, M.J.G. Lee: Phys. Rev. B10, 1213 (1974)
6.18 N.A.W. Holzwarth: Phys. Rev. B11, 3718 (1975)
6.19 P.T. Coleridge: J. Phys. F: Metal Phys. 5, 1317 (1975)
6.20 N.A.W. Holzwarth, M.J.G. Lee: Phys. Rev. B13, 2331 (1976)

6.21 M.J.G. Lee, N.A.W. Holzwarth, P.T. Coleridge: Phys. Rev. B13, 3249 (1976)
6.22 N.A.W. Holzwarth, M.J.G. Lee: Phys. condens. Matter 19, 161 (1975)
6.23 J.D. Eshelby: J. Appl. Phys. 25, 255 (1954)
6.24 P.H. Dederichs, J. Pollmann: Z. Phys. 255, 315 (1972)
6.25 F.J. Blatt: Phys. Rev. 108, 285 (1957)
6.26 W.R. Wampler, T. Schober, B. Lengeler: Phil. Mag. 34, 129 (1976)
6.27 R.R. Bourassa, B. Lengeler: J. Phys. F: Metal Phys. 6, 1405 (1976)
6.28 P. Soven: Phys. Rev. B5, 260 (1972)
6.29 E.A. Stern: Phys. Rev. 168, 730 (1968)

Polariton Theory of Resonance Raman Scattering in Solids

Bernard Bendow

1. Introduction

1.1 Purpose and Scope

Polaritons are the composite quasiparticles formed by the coupling of light with crystalline excitations (see, for example BURSTEIN and DEMARTINI, 1974). Since polaritons are derived from an exact solution of an interaction Hamiltonian, they provide the most physically satisfying basis for the description of a variety of optical effects, among them absorption (HOPFIELD, 1958), luminescence (TAIT and WEIHER, 1969; BENOIT A LA GUILLAUME, 1970), nonlinear processes (OVANDER, 1965), and light scattering (OVANDER, 1962). A good deal of literature has been directed at the problem of resonant light scattering mediated by excitons, and this topic will constitute the main subject of this review as well.

The interest in polariton approaches to resonant Raman scattering (RRS) can be understood from a variety of standpoints. As intimated above, it is very appealing physically to pursue polariton descriptions, since polaritons are the fundamental excitations of light interacting with a crystal. One expects polariton formulations to be more rigorous than those utilizing noninteracting quasiparticles. It is rather intuitive, for example, that in the resonance regime where photon and exciton dispersions cross, a perturbation approach which treats these excitations as uncoupled must break down. Polariton approaches, on the other hand, offer the prospect of an exact treatment of the dispersion induced by the (bilinear) photon-exciton interaction. Other advantages of a polariton formulation include the correct and consistent determination of $\underline{k}$-dependent factors in scattering rates, and the possibility of accounting for multimode excitation effects at boundaries (stemming from a number of polariton branches being degenerate at a particular frequency). A final aspect that is not specifically associated with resonance conditions, but is nonetheless of significant interest, is the incorporation of final-state polariton effects. We shall touch on this subject only briefly here, since our main concern is with the *mediation* of scattering by polaritons, and because it has been covered very adequately in a number of recent reviews (see, for example, BARKER and LOUDON, 1972; MILLS and BURSTEIN, 1974; and CLAUS et al., 1975).

The principal purpose of the present review is to organize and sort the existing literature on the subject of polariton-mediated light scattering, and to critically evaluate the implications and the utility of the polariton approach. We shall especially inquire whether or not polariton predictions differ from those of perturbation theory, and whether or not any predicted polariton effects can in fact be distinguished experimentally. Hopefully, the present review can provide insights that will aid experimentalists as well as theorists in answering questions such as these.

While the prospect of a polariton formulation of RS might seem very attractive indeed, an assortment of difficulties arises in connection with its implementation. These difficulties, which will be explored in some detail in this review, are confined mostly to frequencies in the resonance regime (as one might expect, polariton results simply reduce to those of perturbation theory far from resonance). Stated briefly, these difficulties are associated with properly relating the scattering inside of crystals by polaritons to measurements of photons outside of the crystal. Moreover, depending on the formulation utilized, it is necessary to consider the scattering of damped quasiparticles, for which single well-defined quantum states do not exist, and which are characterized by unphysical group velocities.

The nature and role of these difficulties will be made more explicit further on in the development. We simply note here that we will be describing two characteristically different approaches to the formulation of polariton-mediated RS. In the first, the efficiency is assumed to factorize into appropriate transmission factors (which relate photons outside the crystal to polaritons inside), and into polariton scattering efficiencies within the crystal. The second approach, on the other hand, attempts to treat the entire scattering process as a single unified event, utilizing a combination of quantum and semiclassical techniques. We shall find certain significant differences implied by the two approaches.

Because the physically important properties of the various possible polaritons are closely similar, we find it convenient, for definiteness and simplicity, to restrict the present discussion to scattering mediated by photon-exciton polaritons, with the possible production of final-state photon-phonon polaritons. Extensions to scattering by other polaritons follows directly from information readily available in the literature (see, for example, MILLS and BURSTEIN, 1974).

The reader who lacks a general familiarity with RRS in solids is directed to a variety of previous reviews on the subject, such as those by LOUDON, 1963, 64; MARTIN and FALICOV, 1975; and RICHTER and ZEYHER, 1975, for example. More detailed accounts of original research in the field over the last decade are available in the proceedings of the Light Scattering Conferences (WRIGHT, 1967; BALKANSKI, 1971; BALKANSKI et al.,1976), and the US-USSR Symposium on Light Scattering (BENDOW et al., 1976).

The plan of the paper is as follows: In Section 1.2 we very briefly review certain aspects of the perturbation theory approach to resonant RS. Section 2.1 reviews the fundamentals of polaritons and Section 2.2 introduces the two principal approaches to the formulation of the polariton theory of RRS. The predictions and implications of polariton theory are examined in some detail in Section 3, while Section 4 consists of some concluding remarks on the subject.

1.2 Review of Perturbation Theory

The perturbation approach to RRS and its implications have been described in detail in a wide variety of papers (see, for example, LOUDON, 1963; GANGULY and BIRMAN, 1967; PLATZMAN and TZOAR, 1969; MARTIN, 1971; BARKER and LOUDON, 1972). We shall here direct our attention to just certain salient features which are useful for comparison purposes when discussing the polariton approach to RRS.

In this section and throughout the remainder of this paper we shall employ a simplified model Hamiltonian which will allow us to deduce the principal features of RRS with a minimum of mathematical and notational complexity. Extensions to more general cases will be straightforward from the development.

Let us take for the Hamiltonian H (see, for example, GANGULY and BIRMAN, 1967):

$$
\begin{aligned}
&H = H_o + V; \quad V = V_1 + V_2 \\
&V_1 = V_1^{(a)} + V_1^{(b)} \\
&\bar{H} = H_o + V_1 \\
&H_o = \sum_{\underline{k}} kc a^\dagger_{\underline{k}} a_{\underline{k}} + \sum_{\underline{k}\gamma} E_\gamma(\underline{k}) c^\dagger_{\underline{k}\gamma} c_{\underline{k}} + \sum_{\underline{k}s} \Omega_s(k)\, b^\dagger_{\underline{k}s} b_{\underline{k}s} \, . \\
&V_1^{(a)} = \sum_{\underline{+},\underline{k}\gamma} (\underline{+})\, i|\bar{g}_\gamma|(kc)^{-1/2}\, a^\dagger_{\underline{k}}\, c^\dagger_{\underline{+k}} \qquad (1.1) \\
&V_1^{(b)} = \sum_{\underline{+},\underline{k}} (\underline{+})\omega_p^{\,2}\, k^{-1}\, a^\dagger_{\underline{k}}\, a^\dagger_{\underline{+k}} \\
&V_2 = \sum_{kk'\gamma\gamma's} f_{\gamma\gamma'}(\underline{k},\underline{k}')\, c^\dagger_{\underline{k}'\gamma'} c^\dagger_{\underline{k}\gamma}\, b_{s,\underline{k}-\underline{k}'} + \text{h.c.} \\
&\bar{g}_\gamma = g_\gamma \sqrt{E_\gamma}\,, \quad \omega_p^{\,2} \equiv 4\pi\, N\, e^2/m_e
\end{aligned}
$$

where $(a^\dagger, a)$ are creation-annihilation operators for photons, $(c^\dagger, c)$ for excitons, and $(b^\dagger, b)$ for phonons; $E_\gamma(\underline{k})$ and $\Omega_s(\underline{k})$ are the exciton and phonon energies, respectively ($\hbar = 1$), with $\underline{k}$ the wave vector and γ and s branch indices (we note that γ can be either a discrete or continuous index for the excitons); ω_p is the exciton plasma frequency and N the electron density. H_o is the noninteracting Hamiltonian for the three fields; $V_1^{(a)}$ stems from the p·A interaction between the crystal electrons and the electromagnetic field; $V_1^{(b)}$ comes from the A^2 interaction term; and V_2 is the exciton-phonon interaction. We have limited the Hamiltonian to just a single polarization of light, although the polarizations must of course be reinstated to obtain the tensorial character of the scattering amplitude. For simplicity, we have omitted bilinear phonon-exciton interactions; their omission does not qualitatively change the essential resonance properties of the predicted cross section (see Eqs. 29-31 of GANGULY and BIRMAN, 1967, for example). The $\bar{H}$ defined above, which incorporates all bilinear interactions, is the polariton Hamitonian to be treated in Section 2.

The perturbation picture of a first order RS event is indicated in Fig. 1 and consists of:

a) transmission of the photon ω into the crystal
b) annihilation of the photon with creation of an exciton (via V_1)
c) scattering of the exciton accompanied by creation (Stokes) or annihilation (anti-Stokes) of a phonon (via V_2)
d) annihilation of the scattered exciton and creation of the photon ω' (via V_1)
e) transmission of the photon ω' out of the crystal.

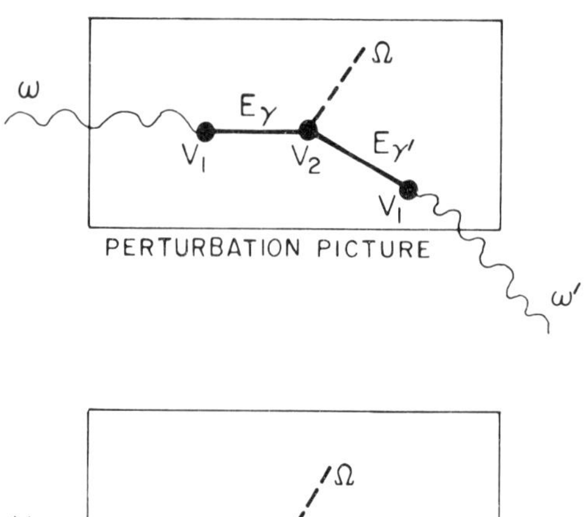

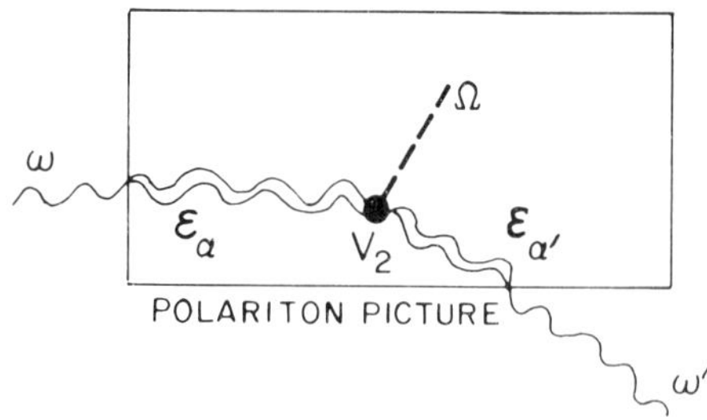

Fig. 1. Perturbation theory and polariton pictures of single-phonon Stokes RS. Single wavy lines: photons; double wavy lines: polaritons; straight solid lines: excitons; dashed lines: phonons

We limit ourselves here to Stokes RS at zero temperature, which is sufficient to demonstrate all the resonance properties we need be concerned with. The probability per unit time for RS is taken to be simply the transition rate from the initial state ψ_i containing one photon ω and no excitons, and the final state ψ_f containing one photon ω', one phonon Ω, and no excitons. According to the usual scattering theory (MESSIAH, 1962) this rate is given by ($\hbar$=1)

$$\frac{dP_{if}}{dt} = 2\pi \sum_f |T_{if}(\omega)|^2 \left[\delta\ \omega-\omega' - \Omega_s(\underline{k})\right] \qquad T_{if} = \langle\psi_f|\hat{T}|\psi_i\rangle \tag{1.2}$$

where $\hat{T}$ is the scattering-operator,

$$\hat{T}(\omega) = V + VGV = V + VG_oV + VG_oVG_oV + \dots \qquad G(\omega) = (\omega-H+i\varepsilon)^{-1},\ G_o(\omega) = (\omega-H_o+i\varepsilon)^{-1}. \tag{1.3}$$

The lowest order contribution to RS from the above expansion must contain two $V_1^{(a)}$ interaction terms (corresponding to incident and scattered photons), and the interaction V_2 (to create or annihilate phonons). We thus require the term $V_1^{(a)}G_oV_2G_oV_1^{(a)}$ in $\hat{T}$, which leads to

$$T_{if} \to \sum_{\gamma\gamma'} \langle i|V_1|\gamma\rangle \frac{1}{\omega-E_\gamma} \langle\gamma|V_2|\gamma's\rangle \frac{1}{\omega-E_\gamma-\Omega_s} \langle\gamma's|V_1|fs\rangle \tag{1.4}$$

where an abbreviated, schematic notation has been employed. Assuming $f(\underline{k},\underline{k}')=f(\underline{k}',\underline{k})$, restricting ourselves to energy-conserving transitions, and taking the two time orderings for creation and annihilation of ω and ω' into account, we obtain ($\hbar$, c = 1)

$$T_{if} = \frac{1}{\sqrt{kk'}} \sum_{\gamma\gamma'} |g_\gamma||g_{\gamma'}|f_{\gamma\gamma'}(k,k')\left[\frac{1}{(\omega-E_\gamma)\ (\omega'-E_{\gamma'})} + \frac{1}{(\omega+E_\gamma)\ (\omega'+E_{\gamma'})}\right]. \tag{1.5}$$

If we now consider a single phonon Ω_o and neglect its dispersion, then

$$\frac{dP_{if}}{dt} \sim k'^2|T_{if}|^2\ \delta(\omega-\omega'-\Omega_o) \sim \frac{k'}{k}\ |R(\underline{k},\underline{k}')|^2 \qquad R(\underline{k},\underline{k}') = \sum_{\gamma\gamma'} \frac{f_{\gamma\gamma'}|g_\gamma||g_{\gamma'}|\ \ (\omega\omega'+E_\gamma E_{\gamma'})}{(\omega^2-E_\gamma^{\ 2})\ (\omega'^2-E_{\gamma'}^{\ 2})} \tag{1.6}$$

where $\omega' = \omega-\Omega_o$. In the present perturbation analysis $\underline{k}'$ and $\underline{k}$ are the wave vectors of the noninteracting photons, so that $k'/k = \omega'/\omega \approx 1$, since $\omega_o \ll \omega,\omega'$. Thus, when the couplings f and g are independent of, or depend only weakly on wave vector, then the resonance behavior is nearly entirely determined by the factor $[(\omega^2-E_\gamma{}^2)(\omega'^2-E_{\gamma'}{}^2)]^{-2}$ above. When γ and γ' are discrete, then a second-order pole arises whenever either $\omega \sim E_\gamma$ or $\omega' \sim E_{\gamma'}$ (incident and scattered resonances); note also that the resonance becomes "doubly strong" if the latter conditions are satisfield simultaneously for some set of variables (see, for example, YU et al., 1973). If the levels γ form a continuum, then the strong resonance associated with discrete states will be suppressed, and replaced instead by an extended regime of enhancement for values of ω and ω' falling into the frequency range spanned by the E_γ (LOUDON, 1963; BENDOW et al., 1970; BENDOW, 1970). For the polariton problem, one is naturally concerned primarily with discrete state resonances.

Several remarks are in order. If the bilinear phonon-exciton term is retained in V, then six terms rather than two contribute to T_{if} (see GANGULY and BIRMAN, 1967, for example), although the resonance behavior is not significantly affected. Another point is that the $f_{\gamma\gamma'}$'s in (1.6) can have different signs for different combinations of (γ,γ'), and thus introduce antiresonances into dP/dt (RALSTON et al., 1970). In fact, the contributions from the discrete and continuous portions of a single exciton band may interfere constructively and destructively in different frequency regimes (BENDOW, 1970). Another consideration of importance is the necessity to include lifetime effects in the energy denominators $(\omega^2-E_\gamma{}^2)^{-2}$, via the prescription $E_\gamma \rightarrow E_\gamma + \Delta_\gamma(\omega) + i\Gamma_\gamma(\omega)$ (BENDOW and BIRMAN, 1971). In this way the obviously unphysical singularities in dP/dt for discrete γ are eliminated. The detailed shape of dP/dt vs ω at resonance is determined in principle by the frequency dependence of $\Gamma(\omega)$, although this dependence may often be neglected in practice. Note that for a continuum of levels the inclusion of an imaginary part in $\omega^2-E_\gamma{}^2$ is especially critical, since this leads to terms describing creation of real (as opposed to virtual) excitons (see, for example, MARTIN, 1974). RSS within a continuum of levels actually represents a mixture of virtual-exciton (Raman-type) processes and absorptions followed by frequency-shifted reemission (resonance fluorescence), all of these being contained in dP/dt when the imaginary part referred to above are retained. Again, these considerations are not particularly germane as far as polariton effects are concerned, although continuum effects do influence polariton dispersion, for example.

The present results, although simplified, should provide an adequate backdrop for discussions of the polariton formulation of RRS.

2. Polaritons and Their Scattering

2.1 Fundamentals of Polaritons

The concept of mixed electromagnetic and crystalline modes appears to have been first explored classically by HUANG, 1951a, b and POULET, 1955, and quantum-mechanically by FANO, 1956. The extension to coupling of light with excitons was carried out by HOPFIELD, 1958 and AGRANOVICH, 1960, both utilizing a second quantized representation. The development given here follows closely that of the latter papers, in a form presented previously by BENDOW, 1970.

Transformation Coefficients and Polariton Dispersion. The Hamiltonian $\bar{H}$ for a single polarization of light interacting with excitons in an isotropic crystal has been given above in (1.1). Since we do not include the bilinear phonon-exciton interaction in H, the phonon portion of H_0 may be ignored in the considerations that follow. It should be remarked that there is absolutely no difficulty formally with including these terms; they are omitted for reasons of convenience alone. $\bar{H}$ defined in this manner constitutes the polariton portion of the Hamiltonian which is central to the present paper.

We search for the normal modes of $\bar{H}$ by defining polariton operators $(A^{\dagger}_{k\gamma}, A_{k\gamma})$ which are linear combinations of the a's and c's, i.e.,

$$A^{\dagger}_{\underline{k},\alpha} = \chi_0(\underline{k})a^{\dagger}_{\underline{k}} + \sum_{\gamma} \chi_{\gamma}(\underline{k})c^{\dagger}_{\underline{k},\gamma} + \phi_0(\underline{k})a_{-\underline{k}} + \sum_{\gamma} \phi_{\gamma}(\underline{k})c_{-\underline{k},\gamma} \tag{2.1}$$

and determine the coefficients (χ,ϕ) which satisfy the equations of motion

$$\left[\bar{H},A^{\dagger}_{i}\right] = \pm\,\omega\, A^{\dagger}_{i} \tag{2.2}$$

This leads, in matrix form, to

$$\begin{bmatrix} kc + \frac{2\omega_{\rho}^{2}}{kc} & \bar{g}_{\gamma}k^{-1/2} & \frac{2\omega_{\rho}^{2}}{kc} & \bar{g}^{*}_{\gamma}k^{-1/2} \\ \bar{g}^{*}_{\gamma}k^{-1/2} & E_{\gamma} & \bar{g}^{*}_{\gamma}k^{-1/2} & 0 \\ -\frac{2\omega_{\rho}^{2}}{kc} & -\bar{g}_{\gamma}k^{-1/2} & -(kc + \frac{2\omega_{\rho}^{2}}{kc}) & -\bar{g}^{*}_{\gamma}k^{-1/2} \\ -\bar{g}_{\gamma}k^{-1/2} & 0 & -\bar{g}_{\gamma}k^{-1/2} & -E_{\gamma} \end{bmatrix} \begin{bmatrix} \chi_0 \\ \chi_{\gamma} \\ \phi_0 \\ \phi_{\gamma} \end{bmatrix} = \omega \begin{bmatrix} \chi_0 \\ \chi_{\gamma} \\ \phi_0 \\ \phi_{\gamma} \end{bmatrix} \tag{2.3}$$

where the summation convention is to be used for the products $g_{\gamma}\,\chi_{\gamma}$ and $g_{\gamma}\,\phi_{\gamma}$, but not otherwise. The determinant of the above matrix minus $I\omega$ yields the dispersion relation determining the polariton modes, namely

$$\left(\frac{kc}{\omega}\right)^2 = 1 + \sum_{\gamma} \frac{4g_{\gamma}^2}{E_{\gamma}^2-\omega^2} \equiv \varepsilon(\underline{k},\omega) \tag{2.4}$$

where the sum rule $\omega_p^2 = \sum_{\gamma}|g_{\gamma}|^2$ has been utilized; ε is the contribution to the dielectric function of the crystal from interaction of light with excitons. To account for a background dielectric constant ε_0, the replacement $c \to c/\varepsilon_0$ and $g \to g/\sqrt{\varepsilon_0}$ should be made above. The manipulation of (2.3) can be shown to lead to Maxwell's equations for the photon amplitude χ_0, which represents the admixture of $(a^{\dagger}_{\underline{k}} + a_{-\underline{k}})$ present in the polariton (see ZEYHER et al., 1974).

We denote the solutions for ω of the dispersion relation $(kc/\omega)^2 = \varepsilon(k\omega)$, i.e., the polariton energies, by $E(k\alpha)$. Then apart from an irrelevant constant, the polariton Hamiltonian becomes

$$\bar{H} = \sum_{k\alpha} E(k\alpha)A^{\dagger}_{k\alpha} A_{k\alpha} \quad , \tag{2.5}$$

i.e., $\bar{H}$ is comprised of a collection of noninteracting (harmonic) modes of energy $E(\underline{k}\alpha)$. The polaritons are related to the noninteracting fields through the coefficients (χ,ϕ), which follow as

$$\begin{aligned}
\chi_{\gamma} &= (\omega-E_{\gamma})^{-1}\, \bar{g}^{*}_{\gamma}\, \Lambda\, \omega^{-1/2}, \quad \phi_{\gamma} = -(\omega+E_{\gamma})^{-1}\, \bar{g}_{\gamma}\, \Lambda\, \omega^{-1/2} \\
\chi_0 &= 1/2\ (kc+\omega)\ (kc\omega)^{-1/2}\Lambda, \quad \phi_0 = 1/2\ (kc-\omega)\ (kc\omega)^{-1/2}\Lambda \\
\Lambda^{-2} &= 1 + \sum_{\gamma} 4E_{\gamma}^2|g_{\gamma}|^2\ (\omega^2 - E_{\gamma}^2)^{-2}.
\end{aligned} \tag{2.6}$$

Physically, the (χ,ϕ) measure the fraction of photon or exciton contained in a given polariton mode, as will be seen more explicitly below. One may similarly obtain the inverse relations expressing a's and c's in terms of A's, but we do not do this here.

As examples of the application of the above formalism, let us consider the polariton dispersion and transformation coefficients for various simplified cases. For a hydrogenic exciton band consisting of discrete levels plus a continuum, one obtains the results for E vs k indicated in Fig. 2. The "repulsion" of the polaritons from the bare quasiparticles in the crossover regions is clearly evident; far from the crossover, on the other hand, the polariton takes on the dispersion characteristic of the bare quasiparticles. The requency regime in which dispersion effects are significant is given by $|\omega^2-E_{\gamma}^2| \lesssim 4g^2/\varepsilon_0$. We note that in addition to all the discrete modes illustrated, all energies in the continuum are eigenvalues of the polariton dispersion relation as well. Turning to the transformation coefficients (χ_0,ϕ_0) which measure the "amount of photon" in the polariton, it is clear from (2.6) that these are large everywhere except when $|\omega^2-E_{\gamma}^2| \lesssim 4g^2/\varepsilon_0$; correspondingly, the exciton coefficients are large only for $|\omega^2-E_{\gamma}^2| \lesssim 4g^2/\varepsilon_0$.

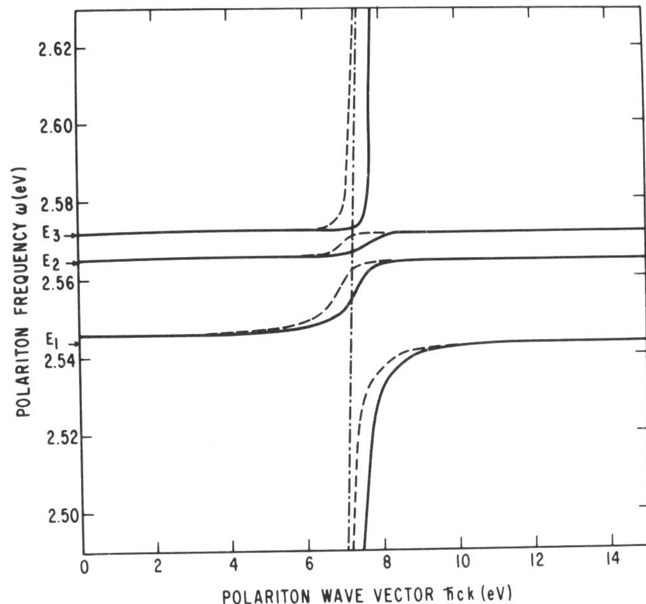

Fig. 2. Calculated polariton frequency vs wave vector for CdS A-exciton parameters. Solid line: full result; dashed line: omitting continuum; dash-dot line: photon line (from BENDOW, 1970)

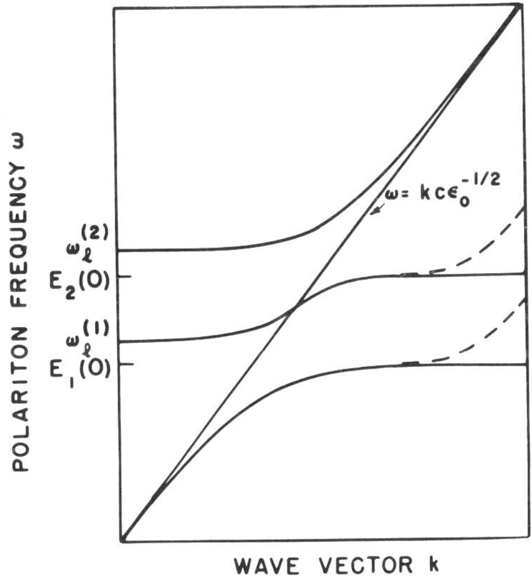

Fig. 3. Polariton dispersion for two nondispersive levels (solid lines) and parabolic levels (dashed lines) (from BENDOW and BIRMAN, 1970)

The modifications of the polariton dispersion for the case of dispersive excitons ($E = E(k)$) is indicated in Fig. 3. The variation of the coefficient $|\chi_1|^2 - |\phi_1|^2$ which measures the "amount of exciton" in the polariton, is illustrated in Fig. 4 for the lower polariton branch. This coefficient is small for $\omega \ll E(0)$, but becomes large and constant for $\omega \gtrsim E(0)$, where the polariton becomes exciton-like in character. Note that in the dispersive exciton case a multiplicity of denegerate polariton modes may coexist for a given frequency ω. For photons incident at such frequencies the appropriate admixture of each of the polaritons excited must be determined in order to calculate the cross section (PEKAR, 1958). The possibility of creating multiple propagating modes, even when just a single dispersive exciton is present, is peculiar to the polariton view of the scattering event. Within a bare exciton approach the photon is assumed to excite all possible exciton states as intermediate states, and the matter of multiple mode excitation (and, as we shall see later, the corresponding necessity for additional boundary conditions) need not be taken into consideration.

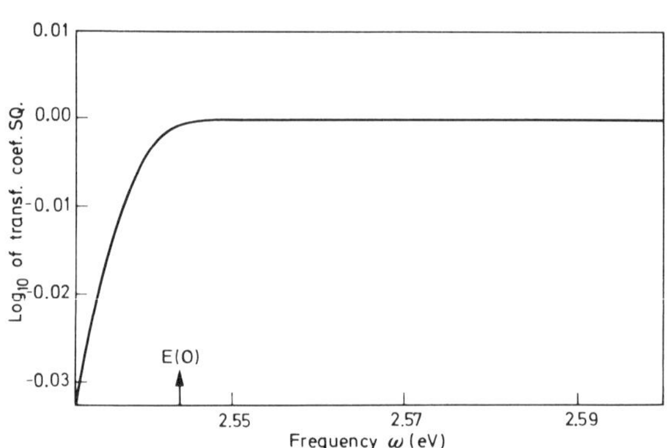

Fig. 4. Log_{10} of exciton polariton transformation coefficient vs frequency for CdS parameters (from BENDOW and BIRMAN, 1970)

Strength Functions. Although calculated scattering amplitudes will, in general, involve quadratic expressions in the (χ,ϕ)'s with arbitrary coefficients, it is nevertheless useful for many purposes to consider simpler functions constructed from the (χ,ϕ)'s. Among these are the "strength functions" (MILLS and BURSTEIN, 1969), which measure the fraction of interacting quasiparticle contained in the polariton. The definition of strength functions is not unique (see, e.g., BENDOW, 1971b); we adopt the following ones because of their relations to certain important matrix elements.

For a single dispersionless exciton E_1 the exciton and photon strengths are given by

$$S_e(\underline{k}\omega) = \frac{|\langle n_\omega+1|c^\dagger_{\underline{k}}+c_{-\underline{k}}|n_\omega\rangle|^2}{n_\omega+1} = \frac{1}{(E_1^2-\omega^2)^2}\,\frac{4g^2E_1\omega}{\varepsilon_0}\,\Lambda^2$$

$$= \frac{\omega}{E_1}\,(|\chi_1|^2 - |\phi_1|^2) \tag{2.7}$$

$$S_R(\underline{k}\omega) = \frac{|\langle n_\omega+1|a^\dagger_{\underline{k}}+a_{-\underline{k}}|n_\omega\rangle|^2}{n_\omega+1} = \Lambda^2 = |\chi_0|^2 - |\phi_0|^2$$

where n_ω is the occupancy of the polariton corresponding to ω. Note that $E_1S_e+\omega S_R=\omega$, which is equivalent to the normalization condition $\sum_i(|\chi_i|^2 - |\phi_i|^2) = 1$. A more detailed exposition of the properties of the S's and their interpretation may be found in MILLS and BURSTEIN, 1974. For our purposes the important point is simply that S_e is maximal at frequencies close to E_1, while S_R displays the opposite trend.

It is perhaps appropriate at this point to note that much of the present development could have been equally well deduced from a Green's function analysis, as performed by MAVROYANNIS, 1967, for example. The advantage of such a formulation is that one may be able to relate scattering efficiencies directly to appropriate Green's functions, or their ancillary functions, without separately calculating transformation coefficients and the like. Nevertheless, we believe that the combination of the equation-of-motion approach presented above with scattering theory for the polaritons provides the most physically transparent approach. On the other hand, Green's functions may prove especially useful with respect to exciton damping effects in polariton scattering, although these have yet to be worked out explicitly within any method.

Damping. If the exciton-phonon interaction V_2 were added to $\bar{H}$, then the polaritons would become damped quasiparticles, since their exciton portion would acquire a finite lifetime. Formally, the effect of the interaction is to modify the polariton eigenenergies through a complex proper self-energy of the form

$$\Pi(\underline{k},\omega) = \Sigma(\underline{k},\omega) + i\Gamma(\underline{k},\omega). \tag{2.8}$$

Σ, the real energy shift, is usually absorbed into E to produce, at least conceptually, the "observed" polariton energy. The effects on Γ, on the other hand, may be much more significant. For example, the polariton dispersion may become substantially altered; sharp-line spectra will acquire finite widths, and possibly detailed structure as well. Unfortunately, the properties of $\Gamma(\underline{k},\omega)$ have not been extensively investigated, and at present explicit results are available only for highly simplified models. Nevertheless, it is possible to obtain insight into some damping-induced phenomena from just certain qualitative considerations, which is the course we shall follow here. For example, it is clear that $\Gamma(\underline{k}\omega)$ is smallest when the polariton is lightlike, since only the exciton portion of the polariton is subject to decay. Correspondingly, Γ is maximal when the polariton is exciton-like; in fact, we expect that Γ is roughly proportional to the exciton strength S_e, and that $\Gamma \to \Gamma_e$, where Γ_e is the bare exciton damping constant, for $S_e \to 1$.

The semiclassical approach may be utilized to incorporate the effects of damping on polariton dispersion. If we add a friction term to the classical equations of motion then the dispersion relation becomes

$$\varepsilon(\underline{k},\omega) = \varepsilon_0 + \sum_{\gamma} \frac{4|g_{\gamma}(\underline{k})|^2}{E_{\gamma}^2(\underline{k}) - \omega^2 - i\omega\Gamma_{\gamma}(\underline{k}\omega)} \tag{2.9}$$

where Γ_γ is the frictional damping constant for exciton γ. A typical dispersion curve for a single level E_1 and constant Γ is illustrated in Fig. 5. One notes that the reflection gap present when $\Gamma=0$ now disappears, and that the polariton dispersion "bends back" at some finite value of $\underline{k}$, in contrast to the undamped case. Detailed curves for various cases are given by AGRANOVICH and GINSBURG, 1966. The strength functions for the present case may be obtained by resorting to a definition of these quantities in terms of crystalline and electromagnetic energy. The result for a single level (MILLS and BURSTEIN, 1974) is obtained by replacing $\varepsilon_0 \to n^2(\omega)$ and $(E^2-\omega^2)^2 \to (E^2-\omega^2)^2 + \omega^2\Gamma^2$ in Λ, in (2.6). Increasing Γ, it turns out, decreases S_e, with S_e remaining less than unity even for $\omega \to E$. It must be remembered, however, that these properties are qualitative in nature. The quantum theory must ultimately be employed to obtain the rigorous $(\underline{k},\omega)$ dependence of Γ, and it would not be surprising if conclusions based on constant Γ approximations turn out to be inaccurate in certain instances. It should also be pointed out that scattering efficiencies do not simply involve strength functions alone, but more general correlators, or response functions, possessing a more complicated structure.

Polariton Velocities. Polariton scattering efficiencies are related to the rate at which polaritons are transported through the crystal. Some of the velocities associated with the polariton are the phase velocity $v_p = E(\underline{k})/k$; group velocity $v_g = \partial E(\underline{k})/\partial k$; and energy velocity v_E, which is defined as the (time-averaged) Poynting vector divided by the energy density. Clearly, the phase velocity is a useful description of wave propagation only when ω depends linearly on k (when the polariton is lightlike). Indeed, for the latter case all three velocities are nearly equal to $v_p = c/n$, where the refractive index n is very slowly varying as a function of frequency. In the photon-exciton crossover region, however, their behavior can differ widely. For dispersionless excitons, for example, all three velocities tend to zero as $\omega \to E_\gamma$, but at different rates. Obviously, none of the velocities exist in the reflection gap for the dispersionless case. However, for a dispersive exciton, the gap is filled and $v_g \to \partial E_\gamma(k)/\partial k$ along the exciton-like portion of the lower polariton branch.

When damping is included, the polariton dispersion may conceivably bend around for $\omega \gtrsim E_\gamma$ (see Fig. 5). Thus, v_g will take on unphysical infinite and negative values. The only velocity of the three retaining its physical significance is v_E, which can be shown to always remain less than c (BRILLOUIN, 1960). For constant damping ($n \to n+i\kappa$)

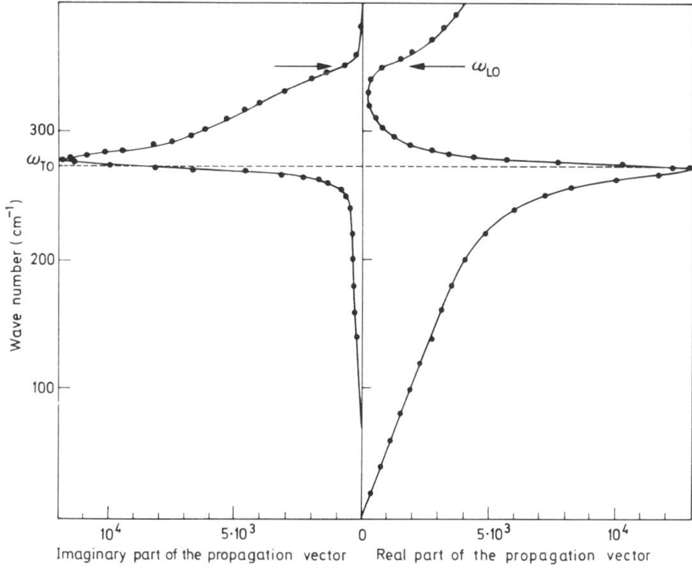

Fig. 5. Frequency vs real and imaginary parts of the wave vector q, for a polariton formed by the TO phonon mode of a zincblende crystal from M. BALKANSKI, in "Optical Properties of Solids", F. Abeles, ed. (N. Holland, Amsterdam, 1972)

$$v_E = c\left[n(\omega)+2\omega\kappa(\omega)/\Gamma\right]^{-1} \tag{2.10}$$

which attains its maximum for $\omega \sim E_\gamma$. A comparison of v_E and v_g for a fairly large value of Γ is indicated in Fig. 6.

2.2 Formalism of Polariton-Mediated Scattering

In this subsection we explore two formulations of polariton-mediated RS, on which the results in Section 3 will be based. A RS event may be pictured in the fashion indicated in Fig. 1b, as opposed to the perturbation picture in Fig. 1a. A central issue in obtaining a polariton formulation is the relation between polariton scattering probabilities and experimentally observed efficiencies, and we therefore focus some attention on this question here. The effects of final-state polaritons (creation of IR-active phonons) does not require fundamental revisions of the usual scattering theory, and therefore we defer this topic entirely until Section 3. In this section we limit the discussion to polariton effects associated with states mediating the scattering.

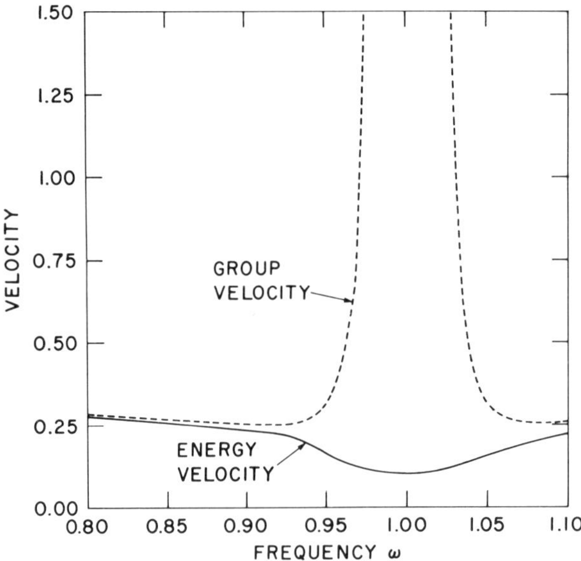

Fig. 6. Comparison of group and energy velocities for a material with $\varepsilon(\omega) = \varepsilon_0 + g^2(E^2 - \omega^2 + i\Gamma)^{-1}$ for $\varepsilon_0 = 10$, $g^2 = 0.2$, $E = 1.0$, and $\Gamma = 0.1$

General Considerations. Why, in general, should one expect scattering calculated utilizing polaritons to differ from that obtained by perturbation theory? A principal reason is that the quasimomenta (wave vectors) of the scattering quasiparticles are no longer constrained to the corresponding external photon momenta, nor do the quasimomenta of the phonons produced in the scattering need to sum to the difference in the external photon momenta. (Of course, quasimomentum must still be conserved within the crystal.) Polaritons for frequencies near resonance are characterized by large $\underline{k}$ values, as opposed to the small values constraining the excitons within the perturbative approach. Variations in polariton momentum may be especially significant in the case of spatial dispersion, and in cases where the coupling functions (photon-exciton, exciton-phonon) are strongly $\underline{k} \equiv (k,\theta,\phi)$ dependent.

The definition of the quantum scattering rate for polaritons inside of an infinite crystal is straightforward and unambiguous. Namely, if the initial and final states are (ψ,ψ') and their energies (E,E'), then the probability per unit time and energy for scattering from $\psi \to \psi'$ via the interaction V is given in lowest order by the Golden rule [retention of just the V term in T in (1.3)]:

$$\frac{d^2P}{dtdE'} = 2\pi|\langle\psi|V|\,\psi'\rangle|^2\delta(E-E')$$

$$\frac{dP}{dt} = \int dE'\,\frac{d^2P}{dtdE'}\,. \qquad (2.11)$$

Typically, we take ψ to be a single polariton and ψ' a polariton plus a phonon. However, it is now necessary to relate this probability for internal polariton scattering to the observed scattering of photons external to the crystal. Not surprisingly, it turns out that any reasonable definition of the observed scattering reduces to the same expression, equivalent to that obtained by perturbation theory, when all polaritons in the scattering are lightlike in nature. As is always the pattern in polariton approaches, differences arise only near to resonance ($\omega \sim E_\gamma$), where the polariton dispersion is substantially altered from that of the bare quasiparticles.

One way to appreciate the effects of polariton dispersion on observed scattering efficiencies is to consider the probability P of scattering during traversal of a photon through a crystal of thickness L. Ignoring attenuation effects,

$$P = \Delta t \frac{dP}{dt} = \frac{L}{v} \frac{dP}{dt} \tag{2.12}$$

where Δt is the polariton traversal time, and v the polariton velocity. We note that P involves the velocity v (v_E in absorbing regions, v_g otherwise); if one could somehow measure the polariton scattering probability dP/dt directly, then the dependence of P on v would be sidestepped. In any case, the above considerations indicate how polariton dispersion may influence observed RS cross sections.

Formulations of Polariton RS. Basically, two approaches have been taken in the literature to define RS cross sections within the polariton framework. The gist of these approaches is the following:

a) *App. I.* The RS process is viewed as initiated by an external photon ω which is either reflected with reflectivity R or transmitted with transmittivity T (R+T=1) upon crossing the crystalline boundary . It it is transmitted, then a unique polariton state ψ is created which, in general, is a linear superposition of all of the n polaritons degenerate with frequency ω [$= E_1(\underline{k}) = E_2(k_2) \ldots = E_n(k_n)$]. In the absence of attenuation, the probability of experimentally obsering a scattering event is proportional to the product of the transmittivity of the initial photon across the entrance boundary of the crystal, the probability of ψ scattering to ψ', and the transmittivity of polariton ψ' across the exit boundary of the crystal. More generally, one must correct for attenuation by reducing the effective propagation length of the crystal for both incident and scattered polaritons; solid angle effects due to the discontinuous change in momentum across the boundary need also be considered (see, e.g., LAX and NELSON, 1974, 1975). Nevertheless, the essential point is that in App. I, RS is viewed as a sequence of quantum events: Transmission of the incident photon, which creates a single unique polariton state or none at all, followed by scattering to a well-defined final state, and finally, transmission leading to detection of the scattered photon.

b) *App. II.* The RS of photons in a crystal is viewed as a single unified process involving refraction, absorption, and scattering, rather than one consisting of a succession of quantum steps. The asymptotic states are pure photons, which become polaritons in the region of space defined by the crystal. The two interactions affecting the asymptotic states are the photon-exciton interaction which produces the polariton, and the exciton-phonon interaction, which scatters them.

It is immediately obvious that App. II is simply an exact description of the scattering event, and that if such a formulation could be successfully implemented, then App. I would be unnecessary. Such an attempt has been carried out in recent years by one group of workers (BRENIG et al., 1972; ZEYHER et al., 1972, 1974). Historically, all of the rest of a rather extensive body of literature on polariton RS has been pursued within the framework of App. I (for example, OVANDER, 1962; MILLS and BURSTEIN, 1969; HOPFIELD, 1969; BENDOW and BIRMAN, 1970). This situation is probably due in part to the relative ease of implementing App. I. Our emphasis in this paper will largely reflect the preponderance of this approach in the literature.

In what follows we discuss the form of the cross section within the two approaches. For simplicity, we restrict attention to single-phonon Stokes scattering at T=0 K, for the case where there is just one polariton from a single branch which corresponds to the incident and scattered photons; moreover, we consider the creation of only IR-inactive phonons. This allows us to focus on the physics of the problem rather than on the many complicating details which often tend to cloud the significant properties of the scattering event.

Approach I. Within this approach one uses the arguments stated above to deduce the probability P_1 that a single photon incident on a crystal is scattered from ω to ω'. The incident photon is transmitted with transmittivity T, and produces a polariton $\underline{k}$ of energy $\omega = E(\underline{k})$, which is subsequently scattered via the interaction V, producing the state $|\underline{k}', \underline{k}-\underline{k}'\rangle$ which contains a polariton of energy $E(\underline{k}')$ and a phonon of energy $\Omega(\underline{k}-\underline{k}')$. Finally, the polariton $E(\underline{k}')$ is transmitted out of the crystal with transmittivity T' to produce the photon ω'. Thus

$$\begin{aligned} \frac{dP_1}{dt} &= TT'\,\frac{dP}{dt} = TT'\,2\pi \sum_{\underline{k}'} |\langle \underline{k}|V|\underline{k}'\rangle|^2 \;\delta\left[E(\underline{k})-E(\underline{k}')+\Omega(\underline{k}-\underline{k}')\right] \\ \frac{d^2P_1}{dt\,d\Xi} &= TT'\,2\pi\, k'^2 |\langle \underline{k}|V|\underline{k}'\rangle|^2 \left[\frac{\partial[E(\bar{\underline{k}})-\Omega(\underline{k}-\bar{\underline{k}})]}{\partial \bar{k}}\right]^{-1}_{\bar{\underline{k}}=\underline{k}'} \end{aligned} \tag{2.13}$$

where $\underline{k}'$ is determined by the energy conservation δ-functions; in general, several such $\underline{k}'$ may exist and a sum over them need be taken. Also, in general, $d\Xi$ is not the observed solid angle; one must use methods such as those described by LAX and NELSON, 1974, to rewrite (2.13) in terms of the observed solid angle $d\bar{\Xi}$ outside of the crystal. If we define $f_{\Xi} = d\Xi/d\bar{\Xi}$ then, formally,

$$d^2P_1/dt\,d\bar{\Xi} = f_{\Xi}\; d^2P_1/dt\,d\Xi \tag{2.14}$$

where $d^2P_1/dtd\Xi$ is now the probability per unit time that a scattered photon is detected within an element of observed solid angle $d\Xi$. This probability is independent of time only as long as the photon (i.e., the polariton) is within the crystal. Therefore one is forced to consider the total probability $dP_1/d\Xi$ that a scattering event occurs during the photon's traversal of the crystal. Neglecting all attenuation effects, then for a crystal of length L,

$$\frac{dP_1}{d\Xi} = \frac{L}{v} \frac{d^2P_1}{dtd\Xi} \tag{2.15}$$

where v is the velocity of the incident polariton beam in the crystal. Since there is no first principles prescription which specifies v uniquely, it must be chosen on physical grounds instead, a matter which we address further on in the development. The differential Raman efficiency, which is defined as the number of scattered photons in solid angle $d\Xi$, relative to the number incident, is clearly also given by $dP_1/d\Xi$. Moreover, under steady-state conditions, $dP_1/d\Xi$ may also be interpreted as the number of photons detected per unit time, relative to the number incident per unit time,

$$\frac{dP_1}{d\Xi} = \frac{N\frac{dP_1}{d\Xi}/\Delta t}{\frac{N}{\Delta t}} . \tag{2.16}$$

It should be noted that $dP_1/d\Xi$ times the number of photons incident per unit time is *not* equivalent to the probability per unit time of polariton scattering.

If primes denote the final state quantities, then for $\Omega(\underline{k}) = \Omega_0$ one obtains directly from (2.13 and 15),

$$\frac{dP_1}{d\Xi} = \frac{2\pi f_\Xi TT'k'^2L}{vv'_g} |\langle\underline{k}|V|\underline{k}'\rangle|^2 . \tag{2.17}$$

HOPFIELD, 1969, obtained an equivalent expression with $v=v_g$ starting from the Born approximation for polariton scattering, i.e., v_g was chosen as the physically appropriate choice for v in (2.15). When attenuation is present, the effective thickness L of the sample is altered. Such corrections have been described by LOUDON, 1964, 1965; for backward scattering from normal incidence, for example, L is replaced by $[\kappa(\omega) + \kappa(\omega')]^{-1}$ (for $\kappa^{-1} \ll L$).

Can dP/dt be measured directly? There is one highly idealized case in which it would be possible in principle, namely, backscattering from a semi-infinite crystal in the absence of attenuation. The polariton would be assumed to propagate until a scattering event occurs, in which case it is detected as a backscattered photon. In this instance, since the crystal is infinitely long, the experimentally observed scattering probability[1] per unit time and solid angle, $d^2\bar{P}_1/dtd\Xi$, is simply equal to the corresponding photon scattering probability. Thus $[\Omega(k) = \Omega_o]$

$$\frac{d^2\bar{P}_1}{dtd\Xi} = \frac{d^2P_1}{dtd\Xi} = \frac{2\pi f_\Xi TT'k'^2}{v_g'} |<k|V|K'>|^2. \tag{2.18}$$

Although a velocity v_g' appears above as a consequence of the density of final states, the observed scattering rate does not depend on the incident polariton velocity v (associated with polariton kinematics) which appears in (2.15 and 17). This result is an artifact of the infinite propagation length available to the polariton, and does not correspond to conditions which would normally be realizable in practice. For example, although very long optical fibers could provide the necessary path length, the finite attenuation which is always present and, in fact, maximal in the resonance regime, reduces the effective path length to a small fraction of the nominal length in even the most favorable of cases.

Approach II. We consider the application of App. II to RS developed in various papers authored by ZEYHER, BIRMAN, BRENIG and TING (BRENIG et al., 1972; ZEYHER et al., 1972, 1974; ZEYHER and BIRMAN, 1974). We shall attempt as much as is possible to omit mathematical details, concentrating instead on just the essence of the development. One begins by writing the differential scattering efficiency for single-phonon Stokes scattering as

$$\begin{aligned} \frac{d^2P}{d\Xi d\omega'} &\propto \frac{k_o'^2}{c} \sum_{\underline{q}} \delta\left[\omega-\omega'-\Omega(\underline{q})\right] |T(\underline{k}_o,\underline{k}_o')|^2 \\ T(\underline{k}_o, \underline{k}_o') &= <\bar{0}|a_{\underline{k}_o} b_{\underline{q}} \hat{T}(\omega) a^\dagger_{\underline{k}_o'} |\bar{0}> \\ \hat{T}(x) &= V + V(x-H+i\varepsilon)^{-1}V \\ V &= V_1 + V_2 \end{aligned} \tag{2.19}$$

[1] Note that one could in principle define an efficiency for this case as the number of scattered photons per unit time divided by the incident photon flux,

$$d\bar{P}/d\Xi = N\frac{d^2\bar{P}_1}{dtd\Xi}/Nc = \frac{1}{c}\frac{d^2\bar{P}_1}{dtd\Xi}$$

which, however, possesses dimensions of inverse length. Since the crystal is considered to be infinitely long, $d\bar{P}/d\Xi$ is physically a less appealing quantity to work with than is $d^2\bar{P}_1/dtd\Xi$.

where $|\bar{0}\rangle$ is the noninteracting vacuum, and $\underline{k}_o$'s represent the bare photon momenta; $\hat{T}$ is the scattering operator. The scattering is then recast in terms of the in and out eigenstates $\psi^{\gtrless}$ which are exact in the photon-exciton interaction V_1. These correspond to incoming and outgoing waves which reduce asymptotically to free photons far from the crystal, but are polaritons inside of the crystal. Defining creation-annihilation operators $A^{\pm}$ corresponding to $\psi^{\gtrless}$, and denoting the interacting vacuum as $|0\rangle$, then one finds

$$T(\underline{k}_o,\underline{k}_o') = \langle\psi^{<}|b_{\underline{q}}V|\psi^{>}\rangle = \langle 0|A_{<}(\underline{k}_o)\; b_{\underline{q}}\; VA_{>}^{\dagger}(\underline{k}_o)|0\rangle. \tag{2.20}$$

Thus the efficiency is proportional to the square of the matrix element of V between $A_{>}^{\dagger}|0\rangle$, which is an incident photon at $t = -\infty$ and a polariton within the crystal, and the final state $b_{\underline{q}}^{\dagger}A_{<}^{\dagger}|0\rangle$, containing a polariton and phonon within the crystal and a pure photon outside the crystal at $t = +\infty$. The vital information regarding propagation across the crystal boundaries and within the crystal must now be contained in the states $A_{\gtrless}^{\dagger}|0\rangle$. Clearly, the crucial element in the present formulation will be the procedure utilized to obtain these states.

It is useful to cast T explicitly in terms of the coefficients $\hat{\alpha}_i$ of the exciton amplitude operators $\hat{\alpha}_i = E_i^{1/2}(c_i + c_i^{\dagger})$; the α's are proportional to χ_i's of Section 2.1 for polaritons within an infinite crystal. Although a set of equations analogous to (2.3) obtains for the polariton coefficients in this case, the boundary conditions and labelling of the modes are altered because the polaritons are restricted to a finite crystal region, and must connect asymptotically to pure photons in the exterior of the crystal. To express the scattering in terms of the α_i's, one must specify the form of V_2, which we take to be similar to that of (1.1), namely

$$V_2 = \sum_{ij\underline{q}} f_{ij}(\underline{q})c_i^{\dagger}c_j b_{\underline{q}}^{\dagger} + \text{h.c.} \tag{2.21}$$

Then

$$T(\underline{k}_o,\underline{k}_o') = \sum_{ij} \alpha_i^{<}(\underline{k}_o')^{*}\, \alpha_j^{>}(\underline{k}_o)\, (E_iE_j)^{-3/2} \times \left[f_{ij}(\underline{q})(E_i+\omega')(E_j+\omega)+f_{ji}(\underline{q})(\omega-E_j)(\omega'-E_i)\right]. \tag{2.22}$$

Note that as before the square of the α's should measure the amount of exciton present at ω and ω'. However, since the α's now correspond to excitations in a spatially inhomogeneous medium (crystal plus vacuum), they cannot be obtained directly in $\underline{k}$-space as was previously the case; the transformation to $\underline{r}$-space yields a set of coupled Maxwell's equations for the amplitudes $\alpha^{\gtrless}$ and the photon amplitude corresponding to the vector potential, which may be solved with full account of the boundary conditions for the problem. If we make the ansatz that the quantum amplitudes α_i may be replaced by their classical analogues in $\underline{r}$-space, then the scattering may be calculated from the $\underline{r}$-space version of (2.3) whence

$$T(\underline{k}_0,\underline{k}_0') = 2\sum_{\gamma\gamma'} f_{\gamma\gamma'}(\underline{q})(E_\gamma E_{\gamma'}+\omega\omega')\,(\omega E_{\gamma'})^{-3/2}\,S_{\gamma\gamma'}(\underline{q})$$
$$S_{\gamma\gamma'}(q) = \int d\underline{r}\; e^{i\underline{q}\cdot\underline{r}}\;\alpha_\gamma^{>}(\underline{r})\;\alpha_{\gamma'}^{<}(\underline{r}). \tag{2.23}$$

It is also useful to note that α is related to the photon amplitude $A(\underline{r})$ by

$$\alpha_\gamma(\underline{r}) \propto \frac{E_\gamma^{\,2}}{\omega}\;\frac{g_\gamma}{E_\gamma^{\,2}-(\omega\underline{+}i\varepsilon)^2}\;A(\underline{r}). \tag{2.24}$$

Perhaps the most striking feature of (2.23 and 24) and one which clearly emphasizes the differences between Apps. I and II, is the behaviour predicted for $\omega\!\to\!E_\gamma$. Since A is normalized to a constant outside of the crystal, the α's, and consequently $T(\underline{k}_0,\underline{k}_0')$, blow up as either ω or $\omega'\!\to\!E_\gamma$. This feature, which is a consequence of the semiclassical determination of the amplitudes in App. II, contrasts strongly with the maximum finite strengths of the exciton amplitudes, and the matrix elements occurring in App. I. While a classical oscillator driven at resonance absorbs energy without limit, in the quantum scattering formulation of App. I a single photon can at most create a single polariton of fixed energy and, therefore, finite amplitude. Within App. I resonance may be considered to be a consequence of the decreased velocity, and therefore longer time, spent by polaritons in the crystal when $\omega\sim E_\gamma$ [see (2.15) and (2.17) for example].

Practical Considerations. Before concluding this section, it is worth pointing out some of the practical difficulties encountered when applying the polariton formulation of RS. One important problem is how to include finite polariton damping which, as indicated previously, may substantially alter the polariton dispersion for $\omega\sim E_\gamma$. More significantly, however, damping may compete with (undamped) polariton dispersive effects in determining the frequency dependence of the scattering near to resonance, and unless both can be handled simultaneously, a detailed interpretation of the observed frequency dependence at resonance will not be possible. Inclusion of damping cannot be accomplished by straightforward modification of the matrix elements, since the eigenstates $|\underline{k}\rangle$ and $|\underline{k}'\rangle$ are no longer well defined in the presence of damping. This difficulty could, in principle, be overcome through a Green's function approach (MAVROYANNIS, 1967; MILLS and BURSTEIN, 1969), by calculating the photon proper self-energy in the presence of polariton damping. However, explicit calculations of this sort have not been reported in the literature.

It thus appears that although actual inclusion of damping effects is relatively straightforward within perturbation approaches (BENDOW and BIRMAN, 1971b; FERRARI et al., 1974), it becomes more complicated within the polariton formalism. The reason is that despite the presence of damping the scattering states are pure photons in perturbation theory; in polariton theory the scattering states are themselves damped to begin with. A qualitative picture of damping effects may be deduced from the modification in the strength functions in this instance, but such predictions are not quantitatively accurate. It is easier to incorporate damping within App. II, if one *assumes* that the principal effect of damping is the modification of the semi-classical amplitudes (as before, one may append appropriate frictional terms to the equations of motion, say). However, it is uncertain whether such a procedure provides a rigorous quantum account of the scattering in the presence of damping.

This concludes our necessarily brief sketch of the formal aspects of RS mediated by polaritons which will serve as a framework for the calculations described in the following section. Readers interested in further elaborations of the subject are directed, for example, to HOPFIELD, 1969; MILLS and BURSTEIN, 1970; and BENDOW, 1971a for App. I; and BRENIG et al., 1972 and ZEYHER et al., 1972, 1974, for App. II.

3. Polariton Theory of the Resonance Raman Effect

This section examines the consequences and predictions of the polariton approaches to RRS delineated in Section 2. Comparisons between polariton predictions, and those of other theories and experiment will be indicated. Sections 3.1 - 3.2 will be restricted to a discussion of polariton-mediated RRS, assuming the quasiparticles which scatter light to be IR inactive. A brief description of the modifications arising when the scattering is caused by (rather than mediated by) polaritons is given in Section 4.

3.1 General Properties of the Scattering Rate

Various features of the frequency dependence of the RS efficiency calculated using App. I (see Section 2.2) are well revealed by inspection of (2.17), which we write as

$$\frac{dP}{d\Xi} \sim \frac{k'^2}{vv_g'} |V_{\underline{k}\underline{k}'}|^2 , \quad V_{\underline{k}\underline{k}'} \equiv \langle \underline{k}|V|\underline{k}'\rangle . \tag{3.1}$$

Note that in contrast to the corresponding photon matrix element arising in third order perturbation theory (see Section 1.2), as long as its dependence on $\underline{k}$ and $\underline{k}'$ is weak, then $V_{\underline{k}\underline{k}'}$ remains bounded for all ω,ω'. $V_{\underline{k}\underline{k}'}$ reaches a maximum when both

$\underline{k},\underline{k}'$ correspond to exciton-like polaritons (i.e., near to resonance) and drops off quickly outside the resonance regime. Very strong resonant enhancement is associated not with variations in $V_{\underline{kk}'}$, but rather with the smallness of v and v_g', and the largeness of k', when ω and $\omega' \to E_\gamma$. While the consequences are ultimately similar to those obtained from a perturbation approach, the physical origin of the effects and their interpretation are clearly different.

Since $V_{\underline{kk}'}$ is the matrix element of an interaction consisting of products of exciton operators taken between polariton states $\underline{k}$ and $\underline{k}'$, one would expect that the RS efficiency might be expressible as an appropriately weighted sum of products of exciton strength functions for $\underline{k}$ and $\underline{k}'$. This is essentially the case for typical choices of V, as demonstrated in detail by MILLS and BURSTEIN, 1969. In fact, if the exciton-phonon coupling is only weakly $\underline{k}$-dependent, then the principal frequency dependence of $V_{\underline{kk}'}$ is

$$V_{\underline{kk}'} \sim S_e(\underline{k}\omega)S_e(\underline{k}'\omega') \tag{3.2}$$

where just single excitons $\underline{k}$ and $\underline{k}'$ have been assumed to correspond to the polaritons $\underline{k}$ and $\underline{k}'$. The properties of S_e have been discussed previously in Section 2; essentially, one obtains a near-Lorentzian line shape near resonance,

$$S \sim \omega\left[(\omega^2-E_1{}^2)^2 + 4E_1{}^2\,|g_1|^2\,\varepsilon_0{}^{-1}\right]^{-1} \tag{3.3}$$

for a single level E_1. Thus $|V_{\underline{kk}'}|^2$ is a product of Lorentzians of width $2E_1|g_1|\varepsilon_0^{-1/2}$, which are displaced by the phonon frequency Ω_0 (for single-phonon scattering). If the coupling coefficient f in V_2 is $\underline{k}$-dependent, than it becomes an additional source of frequency dependence which must be accounted for. In all instances, strong enhancement of $dP/d\Xi$ will arise from the polariton velocities. For example, when exciton damping is ignored $v=v_E=v_g$, so v^{-1} diverges as $(\omega^2-E_1{}^2)^{-3/2}$ as $\omega \to E_1$, for a single dispersionless level.

To examine some effects of exciton damping, let us first assume that all exciton-phonon couplings are much weaker than the exciton-photon interaction, so that to a first approximation we may neglect the effects of damping on the strength functions. Thus, if the interaction V_2 does not involve explicitly $\underline{k}$-dependent couplings, the effects of damping on $V_{\underline{kk}'}$ may be neglected. For example, for typical values of g and Γ, replacement of $(E^2-\omega^2)^2$ by $(E^2-\omega^2)^2 + \Gamma^2$ in (3.3) will have little effect. For $\underline{k}$-dependent coupling, the frequency dependence will be altered due to changes in $k=k(\omega)$ induced by damping. However, in all cases, the principal influence of exciton damping is expected to be through the dynamical factor k'^2/vv_g'. For example, for a dispersionless level, k' will no longer vanish at the longitudinal exciton frequency or increase without limit as $\omega' \to E_\gamma$. The velocity v must be chosen as v_E when damping is present (see Section 2.1); for small Γ, v becomes small but not zero as $\omega \to E_\gamma$ (for a dispersionless level). However, the final state velocity v_g' is not a physically proper velocity when damping is present; this difficulty arises because damping was not incorporated consistently from the start in calculating $dP/d\Xi$. Explicit calculations within App. I which remedy this state of affairs are not available in the literature. Actually, as evident from (2.13), e.g., in the presence of phonon dispersion the factor v_g' in (3.1) will be replaced by $\partial E(\underline{k}')/\partial k' + \partial\Omega_0(\underline{k}-\underline{k}')/\partial k'$; this removes the singularity from $(v_g')^{-1}$ for $\omega' \to E_1$, but does not remedy the unphysical values of $\partial E/\partial k$ which still arise when damping is included. All that can be reliably concluded in this connection within App. I is that damping weakens the photon resonances as $\omega, \omega' \to E_\gamma$ from below. Although it is reasonable to suppose that similar effects occur throughout the resonance regime, the explicit form of the modifications has yet to be established.

When phonon damping Γ_L is included but exciton damping neglected, one obtains

$$\frac{d^2P}{d\Xi dE'} \sim \frac{1}{v} \int dk' k'^2 \frac{\Gamma_L}{\left[\Omega_0{}^2 - (E(k)-E(k'))^2\right]^2 + \Gamma_L{}^2} |V_{\underline{k}\underline{k}'}|^2 , \tag{3.4}$$

i.e., the δ-function in (2.13) is replaced, as usual, by the imaginary part of the interacting phonon Green's function. The double displaced Lorentzians composing $|V_{\underline{k}\underline{k}'}|^2$ are now distributed over a width Γ_L centered about Ω_0, thereby suppressing the frequency dependence of $dP/d\Xi$. Similarly, v_g' is now replaced by a more smoothly varying factor as a function of frequency. Deductions of general applicability are not easily made; the specific form of the frequency dependence must be worked out separately for each particular case.

In a spatially dispersive medium v_g approaches the exciton velocity $\partial E(\underline{k})/\partial k$ in the region where the polariton becomes exciton-like (see Fig. 3). Thus the limiting factor at resonance in the absence of damping is the smallness of the exciton velocity. However, if exciton damping is included, the group velocity once again takes on unphysical characteristics, and cannot be utilized directly in the formula for RS. This state of affairs is not surprising, since once again the damping has not been accounted for consistently to begin with.

To summarize then, within polariton App. I RSS is due to the maximization of the exciton portion of the polariton states near resonance, combined with the smallness of the polariton velocity and/or density of states. The sharpness and strength of the resonance are determined by the size of the exciton-photon coupling, the exciton and phonon dampings, and the exciton and phonon dispersion. All of the latter effects compete to determine the overall frequency dependence of the RS efficiency. At present, explicit results are available only for specialized cases, although qualitative considerations can suggest the possible behavior under more general conditions.

In App. II the Raman efficiency is eventually expressed in terms of semiclassical exciton amplitudes α_i [see (2.23)]. In the absence of spatial dispersion and damping, App. II predicts resonance behavior equivalent to that of perturbation theory, in contrast to App. I. Damping is easily taken into account phenomenologically via modification of the susceptibility, and hence the α_i's. However, App. II does not tell us how to calculate the damping; any prescription for the latter must be formulated separately. The results for the efficiency when temporal damping is included are equivalent, in the simplest instances, to those obtained when the bare exciton approach has been modified to include exciton damping (BENDOW and BIRMAN, 1971b). The principal manifestation of polariton effects in the limit of small damping and weakly dispersive excitons is through the $\underline{k}$-dependence of the interactions and the phonon dispersion. The $\underline{k}$'s involved are no longer determined by conservation of photon momentum, but conservation of the polariton quasimomentum, i.e., $\sum_i(\pm\underline{k}_i) = \underline{k}-\underline{k}'$, where $\underline{k}_i$ are the phonon momenta.

3.2 Calculations for Model Systems

In this subsection we investigate the frequency dependence of resonance RS predicted by polariton approaches for various simplified cases, and compare the results to the corresponding predictions of perturbation theory.

Consider scattering mediated by a series of dispersionless, undamped excitons. In the present discussion we shall be concerned with the "uncorrected" efficiency $dP/d\Xi$ which we define via

$$\frac{dP_1}{d\Xi} = TT'f_\Xi \frac{dP}{d\Xi} \; ; \tag{3.5}$$

the exciton-phonon interaction will be taken to have the form in (1.1). Then the Stokes scattering at T=0 K follows as

$$\frac{dP}{d\Xi} = 2\pi \frac{k'^2}{v_g v_v'} \left| \sum_{\gamma\gamma'} f_{\gamma\gamma'}(\underline{k},\underline{k}')\, U_{\gamma\gamma'}(\underline{k},\underline{k}') \right|^2 \tag{3.6}$$

where

$$U_{\gamma\gamma'}(\underline{k},\underline{k}') \equiv \chi_\gamma^*(\underline{k})\chi_{\gamma'}(\underline{k}') + \phi_\gamma^*(\underline{k})\,\phi_{\gamma'}(\underline{k}'). \tag{3.7}$$

We have above assumed that just single polaritons correspond to (have the same energy as) ω and ω', and have neglected phonon dispersion as well. An especially simple result which reveals many general properties of (3.6) follows when one takes $f_{\gamma\gamma'} = \delta_{\gamma\gamma'} f_\gamma$, whence

$$\frac{dP}{d\Xi} = 2\pi \frac{k'}{k} \left|R(\underline{k},\underline{k}')\right|^2$$

$$R(\underline{k},\underline{k}') \equiv \sum_\gamma \frac{f_\gamma(\underline{k},\underline{k}')\,|g_\gamma|^2(\omega\omega'+E_\gamma{}^2)}{(\omega^2-E_\gamma{}^2)\quad(\omega'^2-E_\gamma{}^2)} \; . \tag{3.8}$$

R is identical, to within a constant, to the Raman amplitude calculated by perturbation theory (see Section 1.2; the equivalence holds for $f_{\gamma\gamma'} \neq \delta_{\gamma\gamma'} f_\gamma$ as well). Denoting perturbation results by "B" and polariton results by "P" then

$$\frac{\left(\frac{dP}{d\Xi}\right)_P}{\left(\frac{dP}{d\Xi}\right)_B} = \frac{k'}{k} \; . \tag{3.9}$$

Essentially, this factor enhances the scattered resonance and suppresses the incident one, as compared to the perturbation theory results.

The factor k'/k is displayed as a function of frequency in Fig. 7 for CdS parameters. The scattering for a single level predicted by (3.8) varies as $(\omega^2-E_1{}^2)^{-3/2}$ as $\omega\to E_1$, and as $(\omega'^2-E_1{}^2)^{-5/2}$ as $\omega'\to E_1$, as opposed to the $(\omega^2-E_1{}^2)^{-2}$ and $(\omega'^2-E_1{}^2)^{-2}$ behaviour predicted by perturbation theory. These same results can be anticipated directly from (3.1); for example, since $V_{\underline{k}\underline{k}'}$ varies slowly compared to v_g, $dP/d\Xi \sim v_g^{-1} \sim (\omega^2-E_1{}^2)^{-3/2}$ as $\omega\to E_1$.

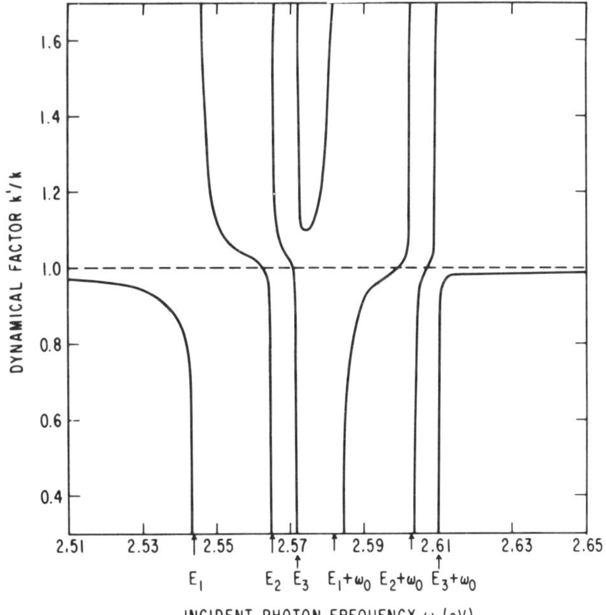

Fig. 7. Dynamical factor k'/k vs photon frequency for CdS A-exciton parameters (from BENDOW, 1970)

Note that the efficiency will be influenced by exactly the same factors involved in calculating $R(\underline{k},\underline{k}')$ within perturbation theory including, e.g., multiple-band effects, discrete-continuum interferences, etc. (see Fig. 8). The frequency dependence of R will be substantially influenced by polariton effects only if R is strongly wave vector dependent. For example, for forbidden phonon scattering where the leading term in $f(\underline{k},\underline{k}') \propto |\underline{k}-\underline{k}'|$, we may expect substantial additional enhancement for ω or $\omega' \to E_\gamma$. Moreover, the scattering kinematics become those of polaritons rather than photons. This situation contrasts with the perturbation approach, where just the difference in (bare) photon momenta, $\underline{k}_0-\underline{k}'_0 \approx 0$ is involved. For the intraband Frohlich scattering experiment reported by MARTIN and DAMEN, 1971, for example, the coupling function f peaks at values of wave vector substantially greater than the values of $|\underline{k}_0-\underline{k}'_0|$ characteristic of the experiment. The latter authors obtained a resonable fit to three widely spaced data points without including polariton corrections. Nevertheless, it is possible for polariton effects to become significant in various instances; a more detailed extension of the measurements closer to resonance may have been desirable to establish their role in the latter case.

Another source of $\underline{k}$-dependence is the photon-exciton coupling $g_\gamma(\underline{k})$. In this instance, both the polariton group velocity and the function $R(\underline{k},\underline{k}')$ will be modified. For the case of quadrupole excitons, for example, $g_\gamma(\underline{k}) \sim \underline{k}$ is the leading dependence of g_γ. One finds (BENDOW, 1971a) that the ratio of quadrupole (Q) to dipole (D) exciton-mediated efficiencies varies as

$$\frac{\left(\frac{dP}{d\Xi}\right)_Q}{\left(\frac{dP}{d\Xi}\right)_D} \sim \frac{k^4 k'^4}{\omega^2 \omega'^2} \tag{3.10}$$

in the regime where the relation $g_\gamma \sim k$ is valid. Far from resonance the above ratio reduces to $k^2 k'^2$, which is the perturbation theory result. The frequency dependence of the Q to D ratio is exhibited for a single level and typical parameters in Fig. 9. Existing experiments on forbidden exciton scattering (BALKANSKI et al., 1972; COMPAANS and CUMMINS, 1973) do not appear to display resonances substantially sharper than those predicted by perturbation theory. However, it should be noted that (3.10) is only valid over a restricted region in $k(\omega)$ and, moreover, does not include damping effects which could be crucial in actual observed cases.

When exciton dispersion is present, then the calculated $dP/d\Xi$ no longer diverges as $\omega, \omega' \to E_\gamma$ as in the dispersionless case. Rather than $v_g \to 0$, $v_g \to v_{exc} \equiv \partial E(k)/\partial k$ near resonance; thus $dP/d\Xi \sim k^{-1}$ for $\omega \to E_\gamma(0)$ and $\sim k'^2/v'_g \sim k'$ for $\omega' \to E_\gamma(0)$, apart from any dependence stemming from $V_{kk'}$. For a single parabolic exciton, for example, $k \sim 2m^*[\omega - E_1(0)]^{1/2}$ for $\omega > E_1(0)$, i.e., in the region where the polariton is exciton-like.

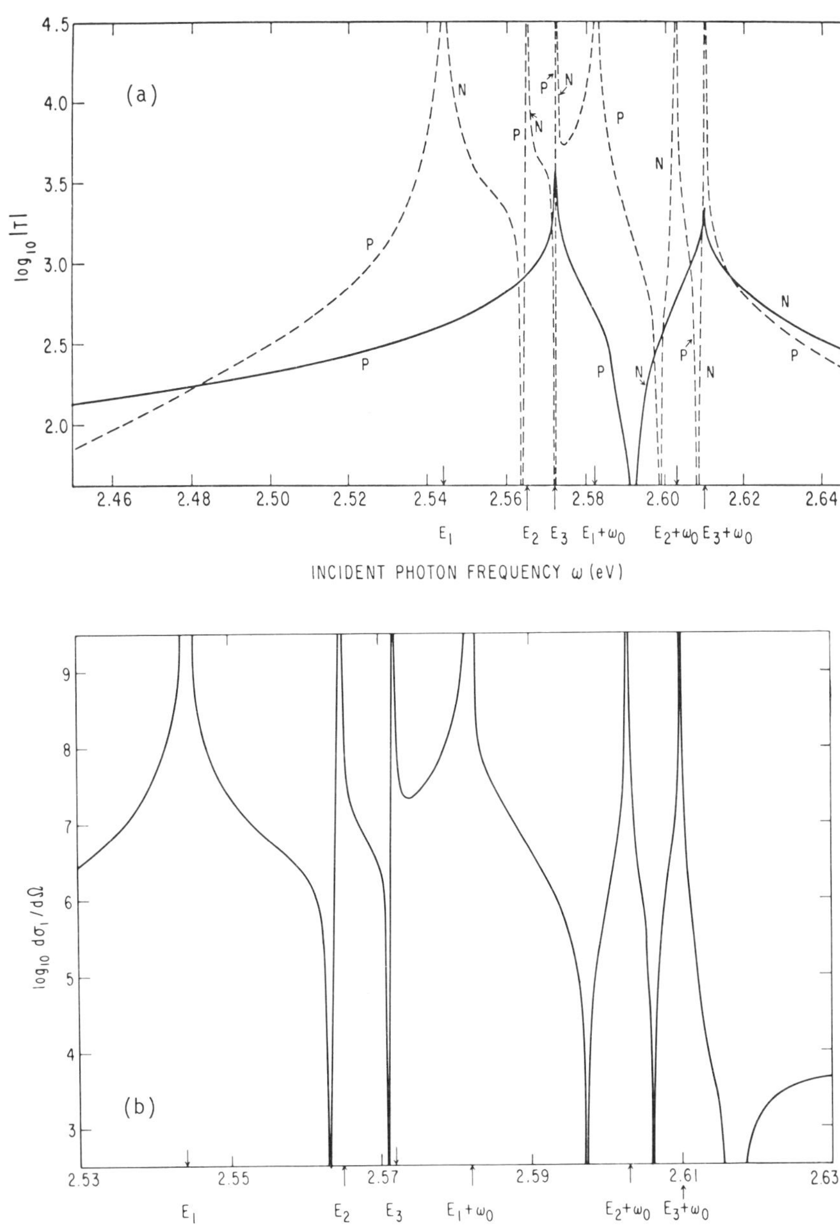

Fig. 8. (a) Contribution to Raman tensor from discrete (broken line) and continuum (solid line) states for A-exciton in CdS. "P" and "N" indicate regions where the contributions are positive and negative, respectively. (b) The polariton RS cross section vs incident frequency ω, calculated from (a)

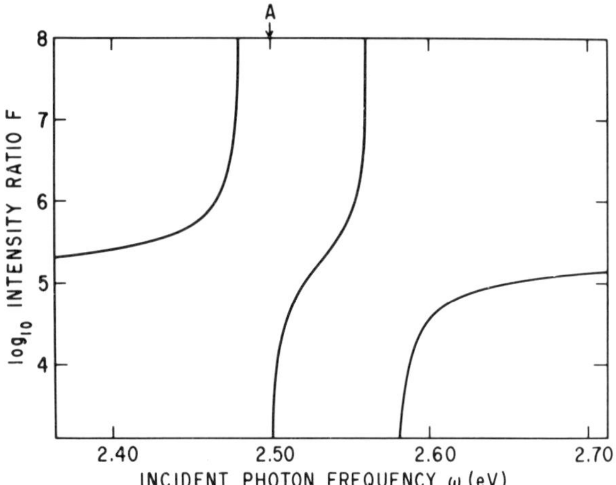

Fig. 9. Log_{10} of quadrupole to dipole intensity ratio F vs incident photon frequency ω. Here the phonon frequency is taken as $\omega_o = 0.08$, the remaining parameters are typical of CdS. "A" indicates the exciton (from BENDOW, 1971)

The resonance enhancement for $\omega, \omega' < E_1(0)$ will be similar to that in the nondispersive case, except that v_g attains its maximum near to E_1 rather than increasing without limit. For zero damping, the in and out resonances occur at the somewhat higher energies E_ℓ and $E_\ell + \Omega_0$ where E_ℓ is the longitudinal exciton, rather than at E_1 and $E_1 + \Omega_0$ (see BENDOW and BIRMAN, 1970). The variation of v_g with frequency for a parabolic exciton is indicated in Fig. 10. Actually, in the case of dispersive excitons, a multiplicity of degenerate polaritons may arise, a circumstance which will be noted in Section 3.3.

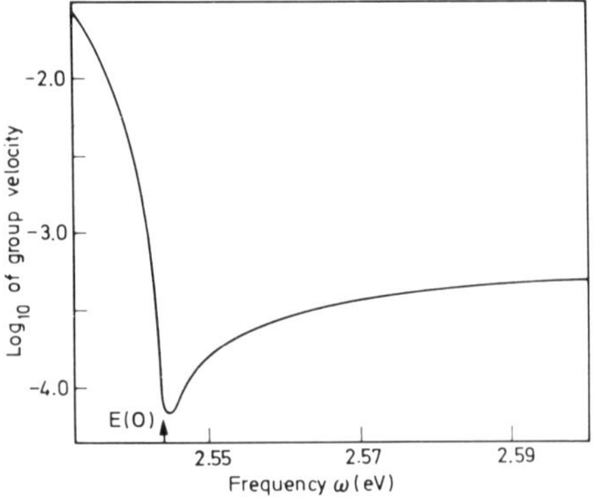

Fig. 10. Log_{10} of group velocity vs frequency for A-exciton in CdS; E(0) is the exciton energy for $\underline{k}=0$

All of the above results have been obtained for cases where exciton and phonon dampings are negligible. Unfortunately, damping may very likely have a major effect on the resonance properties of the efficiency. It is therefore disappointing that explicit calculations have not been carried out within App. I for scattering mediated by damped excitons. In view of the latter, we must necessarily restrict the discussion of damping effects to an extension of the qualitative analysis outlined in Section 3.1. As stated here, unless either the damping is unusually large or $V_{\underline{k}\underline{k}'}$ is strongly $\underline{k}$-dependent, we expect the influence of damping on $V_{\underline{k}\underline{k}'}$ to be relatively minor compared to that on the velocities v and v_g'. In the presence of damping v must be chosen as v_E, so that

$$\frac{dP}{d\Xi} \approx v_E^{-1} = \frac{n+2\omega\kappa/\Gamma}{c} \tag{3.11}$$

as $\omega \to E_\gamma$. The explicit frequency dependence of v_E is somewhat complicated, even for a single dispersionless level. Nevertheless, the general behaviour is rather simple to deduce (see Fig. 6.): far from resonance v_E is similar to the v_g calculated for zero-damping; however, as $\omega \to E_\gamma$, v_E does not tend to zero in the same manner as $v_g(\Gamma=0)$, but bottoms out to some finite value near to E_γ. The smaller the damping, the lower the minimum value of v_E, and hence the greater the resonant enhancement. The trends implied by the latter are similar to those predicted by including damping within a perturbation approach (BENDOW and BIRMAN, 1971b), namely,

$$\left(\frac{dP}{d\Xi}\right)_B \sim \left[(\omega^2 - E_1^2)^2 + \Gamma^2\right]^{-1} \tag{3.12}$$

for $\omega \to E_1$; however, the detailed form of the frequency dependence is clearly diffeent. The damping may also be included phonomenologically in the polariton cross section by changing the various k's in the dynamical factors to the real part of $k(\omega)$, and replacing $E_\gamma \to E_\gamma + i\Gamma_\gamma$ (BENDOW, 1971b). Results for the calculated efficiency are displayed in Fig. 11.

Although parallel arguments to the above cannot be utilized for the scattered photon resonances, it is apparent that damping will eliminate the singularity in the final state polariton density of states (which equals v_g' for $\Gamma = 0$), and that k'^2 will now be bounded as $\omega' \to E_\gamma$. The smoothing out of $v_g'(\omega')$ is analogous to the broadening of harmonic densities of state induced by anharmonic interactions. VERLAN and OVANDER, 1967, in considering scattering for the case where exciton damping is so large that v_g is everywhere less c, utilized v_g in place of v_E to calculate $dP/d\Xi$. However, not only are these conditions not commonly met in practice, but even if they were, it is far from certain that the choice of $v=v_g$ is justified for the large damping case. In cases where $V_{\underline{k}\underline{k}'}$ is strongly $\underline{k}$-dependent (such as for intraband Frohlich coupling), it is evident that damping may substantially suppress the frequency variation exhibited by $V_{\underline{k}\underline{k}'}$. By utilizing the real part of $\underline{k}$ and $\underline{k}'$ in place

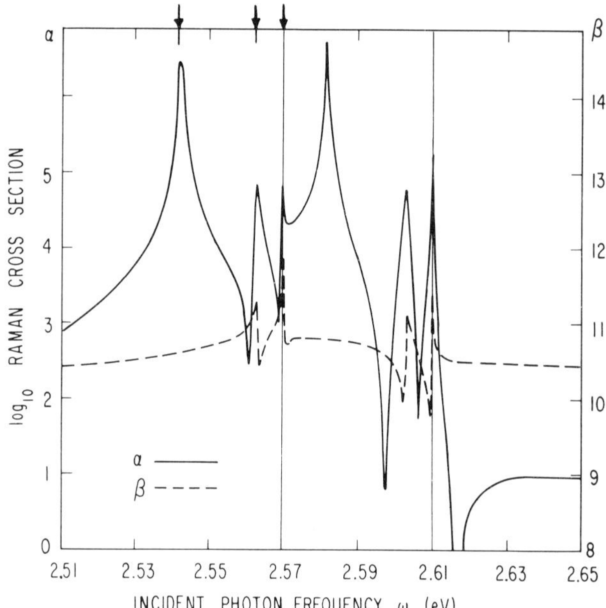

Fig. 11. Polariton Raman cross section with the inclusion of damping, for CdS A-exciton (see Fig. 8). The exciton is represented as a three-level system as indicated by arrows in the figure (all levels for $N \geq 3$ are collapsed into one). The broken line is a calculation for the forbidden exciton case, for which the $N=1$ level does not couple to light

of the zero-damping values, one can obtain a qualitative indication of the modification which will be induced by damping. Expressions obtained by the latter procedures, however, do not necessarily yield quantitatively accurate predictions of the RS efficiency.

As indicated previously, in the absence of spatial dispersion the efficiency predicted by App. II is the same as that obtained in perturbation theory (assuming appropriate transmission and attenuation corrections have been made) for the case where $V_{\underline{k}\underline{k}'}$ depends weakly on $\underline{k}$. For the $\underline{k}$-dependent case, it is merely necessary to replace all bare wave vectors in perturbation theory by their polariton counterparts. Other modifications, some of which will be described later, occur for large values of damping and for scattering in crystals of finite size in special circumstances (ZEYHER et al., 1974). Within App. II the modifications of $dP/d\Xi$ which are induced by exciton damping are usually accounted for via modification of the α_i's, which stem from the particular form of the susceptibility chosen for the crystal. In the presence of damping

$$\chi \sim \sum_i g_i^2(\omega^2 - E_i^2)^{-1} \rightarrow \sum_i g_i^2 \left[\omega^2 - E_i^2 - i\Gamma_i(\omega)\right]^{-1} \tag{3.13}$$

with the eventual result that the factors $(E_\lambda{}^2-\omega^2)$ in (3.8) are replaced by $(E_\lambda{}^2-\omega^2+ i\Gamma)$. This result is essentially the same as one would obtain from perturbation theory when damping has been included. Spatial dispersion effects within App. II will be discussed in Section 3.3.

In Table 1 we contrast the predictions of polariton approaches with those of perturbation theory, for the frequency dependence of $dP/d\Xi$ as $\omega \to E_1$ for a single exciton level, under a variety of assumed conditions. It is to be noted that the assertions by various authors such as BARKER and LOUDON, 1972, that the predictions of perturbation theory and polariton theory for resonant RS are equivalent, are not strictly correct. The formulae used by Barker and Loudon are semiclassical expressions which appear designed to yield the polariton results, and which cannot be obtained from first-principles quantum treatments. Moreover, even granting the validity of their starting expressions, agreement will not be obtained between polariton and perturbation predictions for any other conditions except dipole-exciton mediated, allowed-phonon scattering; predictions for the quadrupole-exciton mediated case, for example, will differ nontrivially [see (3.10)]. And although the differences obtained here between polariton and perturbation approaches may be small from a practical standpoint, they nevertheless represent significant differences in principle, and are thus of interest from a fundamental standpoint.

Table 1. Incident photon resonance behaviour $[\omega \to E_1(0)]$ of Raman efficiency for an isolated exciton level [a]

$D \equiv \omega^2 - E_1{}^2$; table displays results for undamped dipole excitons and allowed-phonon scattering except where otherwise stated.

Case		Perturbation theory	Polariton Approach I	Polariton Approach II
Dispersionless excitons	Allowed phonon	D^{-2}	$D^{-3/2}$	D^{-2}
	[b] Forbidden phonon	D^{-2}	$D^{-5/2}$	D^{-3}
	[b] Quadrupole exciton	D^{-2}	$D^{-7/2}$	D^{-4}
Dispersive exciton		D^{-2}	$\dfrac{\partial E(k)}{\partial k}^{-1}$	[c] $\left[\omega^2-E_1{}^2(k)\right]^{-2}$
Finite damping		$(\omega^2-E_2{}^3+\Gamma^2)^{-2}$	$v_E{}^{-1}$	$(\omega^2-E_1{}^2+\Gamma^2)^{-2}$

[a] Not including reflection or attenuation corrections.
[b] For regime where coupling is linear in wave vector.
[c] $\underline{k}$ is the polariton wave vector determined from $E(k) = \omega$.

An interesting question within App. I to polariton RS is the linewidth of the scattered radiation induced by temporal damping. Since within perturbation theory the only real crystalline quasiparticles involved in the scattering are phonons, the scattered linewidth will simply be that of phonons (LOUDON, 1963). In the polariton picture, however, both a phonon and a polariton constitute the real final state (at least within the crystal), and one might expect the damping to be a sum of phonon (L) and polariton (P) dampings, $\Gamma = \Gamma_L + \Gamma_P$. The polariton damping is large only near resonance (i.e., for $|\omega^2 - E_1{}^2| \gtrsim 2g_1E_1/\varepsilon_0{}^{1/2}$) where it achieves a value close to the exciton damping Γ_E. Combined with the fact that Γ_E and Γ_L are generally of the same order of magnitude, it may be difficult to test the line shape dependence carefully enough for a wide variety of cases to distinguish between the two predictions. There are, of course, many other sources of experimental line shape broadening, be they instrumental or inherent (spatial damping, for example, contributes a linewidth, as will be pointed out in Section 3.3). Existing experiments have not indicated any marked increase in scattered linewidth for $\omega \rightarrow E_\gamma$. For example, the linewidth remains nearly constant for the 1s forbidden exciton resonance in Cu_2O (COMPAANS, 1973), and appears to increase slightly in CuCl (OKA et al., 1973). Recent data of SCHMIDT, 1975, for the TO phonon in ZnSe, where $\Gamma_L \ll \Gamma_P$, suggest that the exciton width does not affect the scattered line shape. It is likely that a more careful theory, which incorporates all damping effects in a rigorous fashion, will be required to yield predictions consistent with these observations.

It is evident from the present discussion that most of the principal features predicted within perturbation and polariton approaches are very similar, such as the nature of the in (ω) and out (ω') resonances, the interferences between discrete and continuum contributions, and the effects of wave vector dependent couplings on the Raman spectrum. Quantitative differences arise because the polariton scattering is governed by the polariton wave vectors $\underline{k}$ and $\underline{k}'$, which become large compared to the photon momenta in the vicinity of resonance. The calculations of BENDOW et al., 1970, indicate that in the case of allowed scattering such differences become pronounced only for large values of exciton-photon coupling g_γ. For wave vector dependent couplings, however, the differences could be more pronounced even for smaller values of g_γ. In all cases, the major effects would occur very close to resonance, where there does not appear to be any existing experiments with sufficient detail and/or accuracy to decide on the differences in exponents n in the energy denominators $(E^2-\omega^2)^{-n}$ predicted within the different approaches. Moreover, the broadening effects of damping and spatial dispersion require a very accurate and detailed determination of frequency dependence in order to distinguish between the various theories. As will be pointed out in Section 3.3, spatial dispersion and finite crystal effects may offer the possibility of detecting the polariton nature of the scattering, although to date only limited data is available. In summary, it seems fair to say that both polariton and perturbation predictions of resonance RS share the same successes as well as failures in their interpretation of experimental results. It is as yet uncertain to what extent the polariton character of the scattering leads to significant, or even unambiguously detectable effects in observed scattering efficiencies.

3.3 Spatial Dispersion and Finite Crystal Effects

As pointed out previously, a feature peculiar to polariton formulations of RS is the possibility of multichannel scattering in the case of spatial dispersion, where a number of degenerate polaritons exist in the incident and/or scattered channels. For example, the four possible channels for scattering via a single dispersive exciton are indicated in Fig. 12. Aspects of single branch scattering in the presence of spatial dispersion were discussed in Section 3.2; in this section we explore multichannel scattering effects. Brief consideration will also be given to finite crystal size and spatial damping effects in polariton RS.

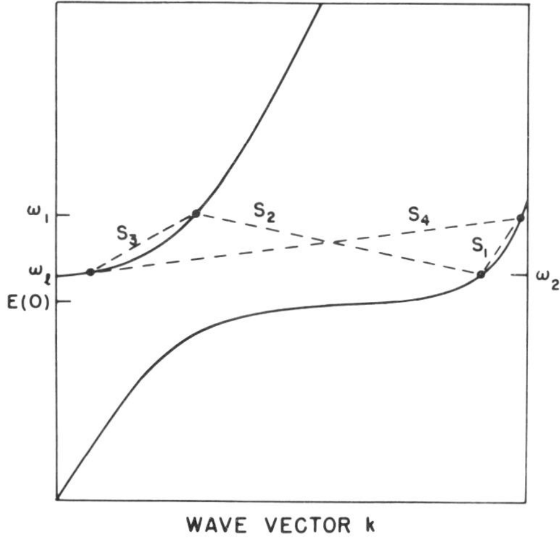

Fig. 12. Four scattering channels for the A-exciton in CdS (from BENDOW and BIRMAN, 1970)

We begin with a description of multichannel scattering within App. I. In the absence of damping, an incident photon ω will create an admixture of N degenerate polaritons $\varepsilon_i(\underline{k}_i) = \omega$, i = 1, 2, ... N. The initial polariton state ψ becomes a linear combination of degenerate polariton states ψ_i, i.e., $\psi = \Sigma_i d_i \psi_i$ with the coefficients d_i determined by appropriate boundary conditions (bc). It is apparent that additional boundary conditions (abc) besides the usual Maxwell or Fresnel relations will be required to uniquely specifiy the d_i's (see, e.g. PEKAR, 1958; AGRANOVICH and GINSBURG, 1966). The nature of the abc has been the subject of some controversy over the years, with rigorous solutions available only in model cases (ZEYHER et al., 1971). Recently, bc's which seem independent of the detailed properties of the surface potential have been proposed for the center-of-mass motion of the exciton (SEIN, 1969, 1970; BIRMAN and SEIN, 1972; AGRAWAL et al., 1971; MARADUDIN and MILLS, 1973). The derivation of these abc actually assume that exciton reflection at the boundary can be neglected (ZEYHER et al., 1972). From the experimental standpoint, most attempts to discern effects of multiple wave excitation in optical spectra appear to be inconclusive. For example, damping and spatial dispersion both exert similar influences on reflection spectra (see SKETTRUP and BALSEV, 1970), so it is difficult to differentiate between them. Observations of interference between multiple polariton waves has been claimed by various investigators (BRODIN and PEKAR, 1960; GORBAN and TIMOFEEV, 1962), with the recent work of KISELEV et al., 1974, being perhaps the most convincing. Rather then delve more deeply into the theoretical and experimental aspects of the abc and multimode excitation problems, we refer the interested reader to the extensive literature on the subject, some of which has been referenced above. We shall here rather be concerned with the general implications of multimode excitation for polariton-mediated RS.

While we expect that multichannel effects will be manifested in principle in the RS efficiency, two factors will strongly influence their role in practice. These are the relative excitation efficiencies $|d_i|^2$, and the relative values of the mode damping functions Γ_i. The contribution of a given channel will be substantially suppressed for large Γ_i and small $|d_i|^2$. For example, exciton-like polaritons will be damped much more strongly than photon-like polaritons (typically by a few orders of magnitude).

The RS efficiency formulated within App. I is not straightforwardly generalizable to the case of multibranch scattering. Note that the only quantity describing the scattering which remains well-defined is the quantum probability for transitions to a final state ψ', which is given as usual by the Golden Rule,

$$\frac{dP}{dt} = 2\pi \int dE_{\psi'} \; |\langle \sum_i d_i \psi_i | V | \psi' \rangle|^2 \; \delta(\omega - E_{\psi'}) . \tag{3.14}$$

In order to deduce the observed scattering efficiency, one must specify the rate at which energy is transported in the polariton state $\psi = \Sigma_i d_i \psi_i$. Since the different portions of the wave propagate with different velocities, it is not obvious whether, or how, a single physically meaningful velocity characterizing the wave can be defined. Moreover, it is unclear how to properly account for attenuation corrections, since each polariton is characterized by a different damping function. These questions can be simply resolved only if interference between channels can be neglected. In this instance the efficiency reduces to just a sum of individual channel efficiencies,

$$\frac{dP}{d\Xi} = 2\pi(1-R) \sum_{ij} T_j'|d_i|^2 \frac{k_j^2}{v(i)v(j)} |\langle \underline{k}_i|V|\underline{k}_j'\rangle|^2 \equiv \sum_{ij} \frac{dP_{ij}}{d\Xi} \tag{3.15}$$

where R is the incident photon reflectivity and T_j the transmission coefficient of polariton j. To obtain R, T_j, and d_i above, one must first solve the multimode boundary problem, as discussed above. The only existing multichannel calculations within App. I are those of BENDOW and BIRMAN, 1970, for CdS. Consider the individual-channel, unweighted efficiences defined by

$$\frac{dP_{ij}}{d\Xi} = \frac{k_j^2}{v_g(i)v_g(j)} |\langle \underline{k}_i|V|\underline{k}_j\rangle|^2 ; \tag{3.16}$$

the calculation of the $dP/d\Xi$'s for the four channels in Fig. 1 are illustrated in Fig. 13a; the final efficiency incorporating the d's and T's is indicated in Fig. 13b. Below E_ℓ only S_1 (purely lower branch) scattering can contribute to the

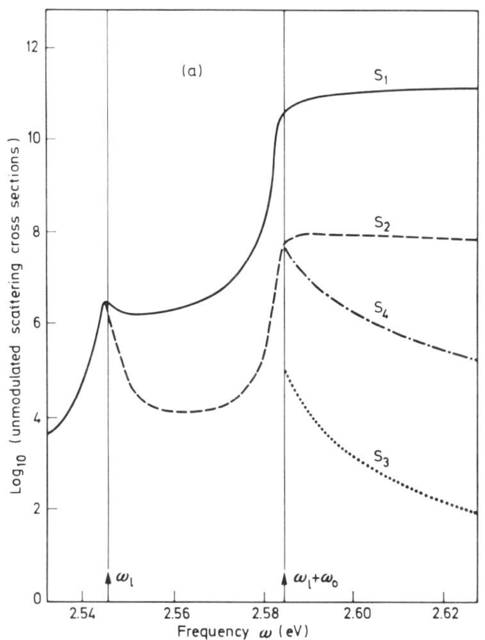

Fig. 13a. $\log_{10}$ of unweighted efficiencies dP_{ij}/dE, vs frequency for the CdS A-exciton, for the four channels indicated in Fig. 10

Stokes efficiency. We note that for $E_\ell<\omega<E_\ell+\omega_0$, S_2 dominates, while for $\omega>E_\ell+\omega_0$, S_3 (purely upper branch) dominates. Typically (interference effects aside), one expects identical trends in the presence of small, but finite damping, *unless* those channels which were dominant for $\Gamma = 0$ turn out to be much more strongly damped

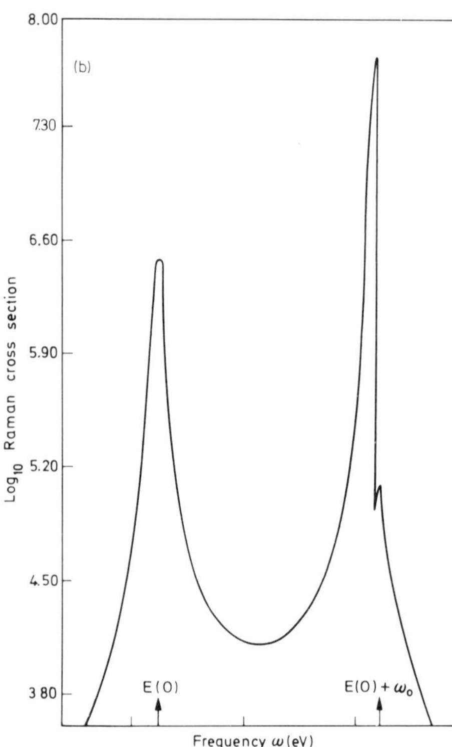

Fig. 13b. Log_{10} of the final Raman efficiency vs frequency calculated from the dP_{ij}/dE's (from BENDOW and BIRMAN, 1970)

for $\Gamma \neq 0$ (dominated by exciton-like polaritons). The latter is usually *not* the case in typical instances.

We note that the multichannel calculations of BENDOW and BIRMAN, 1970, predict efficiencies which are very similar to those obtained by single channel theory (i.e., ignoring spatial dispersion), and thus perturbation theory, for typical parameters, such as in CdS. Greater differences between various calculations are shown to occur for larger photon-exciton couplings. It is doubtful that any of the commonly investigated semiconductors are characterized by sufficiently large couplings to easily differentiate between the various predictions.

App. II is especially well suited to tackle multimode scattering problems, since the principal difficulties surrounding attentuation and propagation of the polariton waves which trouble App. I are, in fact, absent from App. II. In the vicinity of a single level the efficiency for backscattering from a semi-infinite crystal takes the following approximate form (ZEYHER et al., 1974)

$$\frac{dP}{d\Xi} \propto \sum_{ij=1}^{2} A_i A_j \frac{\mathrm{Im}(k_i + k_j + 2k')}{\mathrm{Re}(k_i - k_j)^2 + \mathrm{Im}(k_i + k_j + 2k')^2} D_i(\omega) D_j^*(\omega)$$

$$D_i(\omega) = |g_1|^2 \frac{\omega\omega' + E_1^2}{\left[E_1^2(k_i) - \omega^2\right]\left[E_1^2(k') - \omega'^2\right]} \qquad (3.17)$$

where k' denotes the final state polariton, and $\underline{q} \approx 0$ has been assumed. The field amplitudes A_i are chosen to provide the appropriate admixture of polariton modes obeying the bc's. We have above also incorporated the effects of spatial damping stemming from $S(\underline{q})$ in (2.23). In general, the interference terms ($i \neq j$) are small, since usually $\mathrm{Re}(\underline{k}_1 - \underline{k}_2) \gg \mathrm{Im}(k_1 + k_2)$. The diagonal terms ($i = j$) correspond to intra- or interbranch scattering (depending on $\underline{k}'$); their relative magnitude depends on the abc utilized to determine the A_i's. ZEYHER et al., 1974, found that in frequency regions where both polaritons can propagate, then for equal skin depth the two diagonal terms contribute equally to $dP/d\Xi$. Nevertheless, the skin depth is usually much greater for the phonon-like branch, so its contribution is generally favored (except, perhaps, in very thin samples with $L \cdot \mathrm{Im}k \ll 1$). Note that transmission and attenuation factors do not separate out from $dP/d\Xi$ as they do in simplified cases in the absence of spatial dispersion. Computations of $dP/d\Xi$ vs ω in the vicinity of E_1 for CdS parameters are indicated in Fig. 14. Note that the sharp resonance peak obtained in the absence of spatial dispersion is strongly suppressed in its presence. It is found that for the latter case the extra-channel contributions have a negligible effect on $dP/d\Xi$. Differences in $dP/d\Xi$ with and without spatial dispersion are primarily a result of differences in the polariton dynamics (dispersion) in the two cases. The results in Fig. 14 are not easily compared to the App. I results which were displayed in Fig. 11, for example, since attenuation effects have

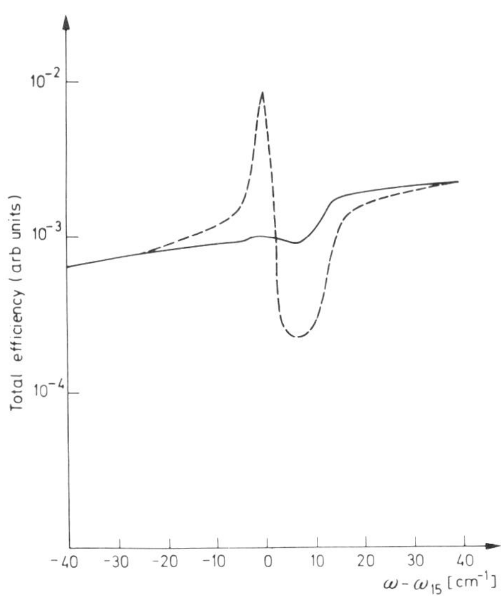

Fig. 14. Sum of both polariton branch contributions to the cross section for first-order allowed scattering. Solid line: calculated with spatial dispersion; dashed line: calculated without spatial dispersion. Corrections for absorption and reflection are included (from ZEYHER et al., 1974)

not been included in the latter. Nevertheless, both approaches appear to agree that multichannel effects typically do not have a significant effect on the frequency dependence of the RS efficiency.

Multimode scattering calculations have been pursued by BRENIG et al., 1972 for the case of Brillouin scattering in resonance with a discrete dispersive exciton level. The principal conclusion is that an octet of lines, rather than the usual

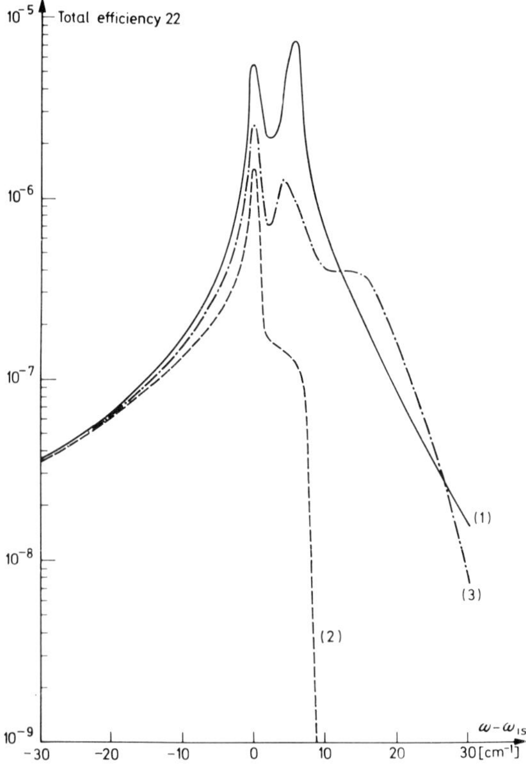

Fig. 15. Lower branch Brillouin scattering efficiency vs frequency calculated for three different abc (from BRENIG et al., 1972). The three abc's take the form

(1) $(P_1-\chi_o E_1) + (P_2-\chi_o E_2) = 0$

(2) $n_1(P_1-\chi_o E_1) + n_2(P_2-\chi_o E_2) = 0$

(3) $$\frac{P_1}{(n_1-n_e)(n_1^2-1)} + \frac{P_2}{(n_2-n_e)(n_2^2-1)} = 0$$

where χ_o is the background susceptibility; n_1, n_2 and n_e are the refractive indices corresponding to the polaritons and the noninteracting exciton; and the P_i's are the polarizabilities of the polaritons doublet, should be observed. Determination of the width and shift of the lines permits, in principle, a complete determination of the dispersions of the various polariton branches. An interesting feature of the calculation is that the relative intensities calculated for the Brillouin lines differ quantitatively depending on the abc's utilized, so that resonance Brillouin scattering measurements may aid in resolving the abc controversy. Computed results for the resonance behaviour are indicated in Fig. 15. Recent experiments in GaAs (ULBRICH and WEISBUCH, 1977) and in CdS (WINTERLING and KOTELES, 1977) have corroborated the prediction of a multiplet of lines at frequencies consistent with scattering from multiple channels. In neither instance were the measurements sufficiently detailed to allow quantitative comparison with theory, and to thus distinguish between the various abc's. And although the GaAs data appear to conform well to the overall prediction of BRENIG et al., 1972, the lines appearing for $\omega > E_\ell$ in the CdS study do not appear to be consistent with these predictions.

App. II suggests that in situations characterized by spatial inhomogeneity, including scattering in thin samples, it may be necessary to take explicit account of boundary effects in calculating efficiencies. Within App. II this is accomplished by solving the (inhomogeneous) Maxwell equations for the polaritons subject to appropriate bc. As noted previously, the factorization of the efficiency into an uncorrected term times transmission and attenuation factors does not obtain in general. To this time, the detailed effects of spatial inhomogeneity and bc on RS have been worked out only in a limited number of specialized cases, from which it is difficult to draw conclusions of a general applicability. One effect of a general nature, however, is the spread in phonon momentum $\underline{q}$ induced by absorption (BRENIG et al., 1972, DRESSELHAUS and PINE, 1975; DERVISCH and LOUDON, 1976). If the absorption is not exceedingly large $[\mathrm{Im}(k) \ll \mathrm{Re}(k)]$ then the principal influence of the distribution in $\underline{q}$-values will be to contribute a line shape to the scattered radiation. The nature of the distribution may be inferred from an inspection of $S(\underline{q})$ in (2.23). In the presence of absorption $A(\underline{r}) \sim \exp\left[i(k_1 + ik_2)\cdot r\right]$ for both incident and scattered waves; employing (2.24), one eventually finds that S leads to a factor proportional to

$$\frac{\mathrm{Im}\,(k' + k)}{\mathrm{Re}(k' - k - q)^2 + \mathrm{Im}(k' + k)^2} \tag{3.18}$$

in $dP/d\Xi$. This result must be modified for very strong absorption, such as in metals, for example, and may also depend on phonon boundary conditions (see MILLS et al., 1968; DRESSELHAUS and PINE, 1975; DERVISCH and LOUDON, 1976). Although detailed investigations are lacking, it appears that the existence of a momentum line shape will not, under usual conditions, significantly affect the resonance properties of $dP/d\Xi$. Cases involving strongly $\underline{k}$-dependent couplings may, however, provide exceptions.

In summary, while polariton approaches predict the existence of a variety of novel features in scattering from spatially inhomogeneous media, only a few isolated cases have been identifies in which the effects appear clearly capable of detection. The one case where the experiments have corroborated the predicted effects is with respect to resonant Brillouin scattering. Existing data are far from adequate to corroborate the detailed nature of the frequency dependence of the RS efficiency. The predictions of the existing theories and their companion experiments will have to be pushed considerably further before the existence and/or nature of various phenomena discussed here can be appropriately ascertained.

3.4 Scattering by Polaritons

When the quasiparticle responsible for scattering couples to light directly (e.g., TO phonons), then a polariton formulation of the scattered spectrum becomes appropriate. The line shape will no longer be characteristic of the bare quasiparticle but will take on a polariton character instead. Moreover, the scattering may consist of contributions stemming from both the bare quasiparticle field as well as from the electromagnetic field associated with the polariton. For scattering by TO phonons, for example, we may write the change in the electronic susceptibility in the schematic form

$$\Delta\chi = au + bE \tag{3.19}$$

where E is the electric field; here a is the usual atomic displacement coefficient, while b is the electrooptic coefficient (see BARKER and LOUDON, 1977 for example). The appearance of the two terms in $\Delta\chi$ creates a possibility for resonant cancellations in the scattered spectrum. In general, a and b will depend on both ω and ω', which will result in resonance properties similar to those described heretofore. We limit the present discussion principally to final-state effects in the nonresonant regime, although some brief comments regarding changes associated with resonance conditions will be made. The present section has been included mainly as a matter of interest, but also for reasons of completeness and perspective. An extensive literature exists on the subject (for example, BENSON and MILLS, 1970; GIALLORENZI, 1971; BARKER and LOUDON, 1972), to which the interested reader is directed for further details. Our brief development here will follow closely that of Benson and Mills (BM).

Consider, for simplicity, zero damping and a case where single a and b coefficients describe the scattering. The E-field may then be related to the displacement u in the standard fashion (see BORN and HUANG, 1956, Ch. II, for example); then one obtains for the change in susceptibility $\Delta\chi$

$$\Delta\chi = Z\mu, \quad Z = a + \frac{\mu}{e^*} b(\omega_{TO}^2 - \Delta\omega^2) \tag{3.20}$$

where $\Delta\omega$ is the frequency shift, μ is the reduced mass, e^* the effective transverse charge, and a and b have been defined above. Under nonresonant conditions $dP/d\Xi$ depends on just the shifts $\Delta\omega$ and $\Delta\underline{k}$, and is proportional to the frequency transform of the susceptibility correlator, i.e.,

$$\frac{dP}{d\Xi} \propto \langle\Delta\chi(t)\,\Delta\chi(0)\rangle_{\Delta\omega} = |Z|^2\langle u(t)u(0)\rangle_{\Delta\omega} \propto |Z|^2 S_L(\Delta\omega) \tag{3.21}$$

where S_L is the phonon strength function (see Section 2.1). For large angle scattering $\Delta\omega\approx\Omega_{TO}$, $S_L \approx 1$, and the scattering becomes similar to that in the absence of polariton effects. For smaller angles $dP/d\Xi$ will display constructive or destructive interference, with the nature and extent depending on the sign and size of the ratio ae^*/b (constructive for positive sign, destructive for negative). Interference effects of this type appear to have been first reported by FAUST and HENRY, 1966, for experiments on GaP. The properties of the Raman spectrum for GaP have been analyzed in some detail by BARKER and LOUDON, 1972, (see their Fig. 7, for example). The analogous interference effect, as observed in ZnSe is illustrated in Fig. 16.

When phonon damping is present, (3.20) and (21) need to be generalized by first writing $dP/d\Xi$ as a linear combination of the four correlators $\langle u(t)u(0)\rangle_{\Delta\omega}$, $\langle E(t)u(0)\rangle_{\Delta\omega}$, $\langle u(t)E(0)\rangle_{\Delta\omega}$ and $\langle E(t)E(0)\rangle_{\Delta\omega}$. These functions are proportional to the imaginary part of the corresponding Green's functions. For the present case where a bilinear interaction describes the polariton, the equations of motion for the set of coupled Green's functions can be solved exactly, as detailed in MB, for example. We here merely state the results corresponding to the generalization of (3.21) to nonzero damping:

$$\frac{dP}{d\Xi} \propto \left|a + \frac{4\pi e^* N\omega^2}{k^2c^2 - \varepsilon_0\omega^2}\, b\right|^2 C(\Delta\underline{k},\Delta\omega) \tag{3.22}$$

$$C(\Delta\underline{k}, \Delta\omega) = \int d\underline{r}\, dt\, e^{i(\underline{k}\cdot\underline{r}-\omega t)} \langle u(\underline{r},t)u(0,0)\rangle$$

where C is the displacement correlator of the lattice in $(\underline{k},\omega)$ space. For a TO phonon mode one obtains (E_1 and E_2 are here phonon-photon polaritons)

$$C(\underline{k},\omega) \propto \frac{\Omega_{TO}\Gamma(\omega)\,(\omega^2 - k^2c^2\varepsilon_0^{-1})}{[\omega^2 - E_1^2(k)]^2\,[\omega^2 - E_2^2(\underline{k})]^2 + 4\Omega_{TO}^2\Gamma^2(\omega)\,(\omega^2 - k^2c^2\varepsilon_0^{-1})} \,. \tag{3.23}$$

Between them (3.22) and (23) embody both polariton effects arising through $C(\underline{k}\omega)$ and those associated with interference between atomic displacement (a) and electro-optic (b) induced scattering mechanisms. The typical variation in scattered linewidth with angle is highly parameter dependent. Results for ZnSe are indicated in

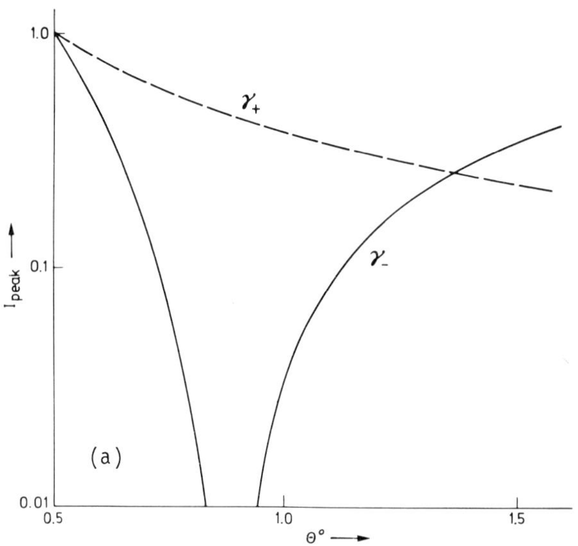

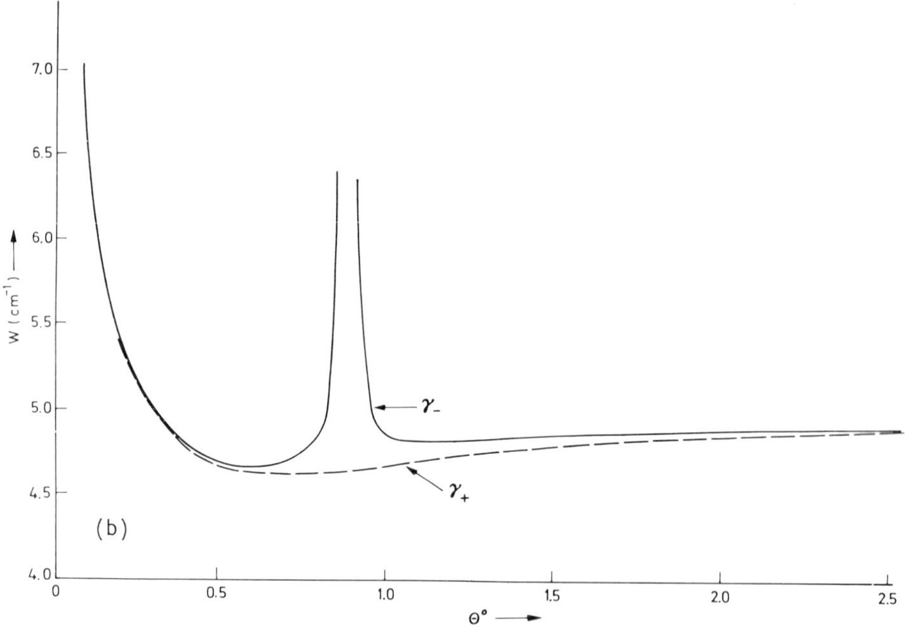

Fig. 16. (a) Dependence of peak Raman intensity on scattering angle, for parameters appropriate to ZnSe (γ^-). (b) Intrinsic full width at half maximum of the Raman line in ZnSe. γ^+ indicates result when the sign of b/a is changed (from BENSON and MILLS, 1970)

Fig. 16b, and display a substantial increase in linewidth for angular values corresponding to cancellation, as would be expected. For one particular configuration in ZnO, on the other hand, BM found a relatively slow variation for the width. MB also found, interestingly enough, that even for fairly large values of damping the peak position in the scattered spectrum is determined very nearly by the undamped polariton dispersion relation. For large angles one is outside the polariton regime ($\Delta\omega^2 << c^2 \Delta k^2 \varepsilon_0^{-1}$) and $dp/d\Xi$ reduces to

$$\frac{dP}{d\Xi} \rightarrow a^2 \frac{\Gamma}{(\omega^2 - \Omega_{TO}^2)^2 + 4\Omega_{TO}^2 \Gamma^2} \tag{3.24}$$

which is formally the same as for scattering by nonpolar phonons. Thus polariton effects are manifested principally in small angle (nearly forward) scattering.

Although the assumption that $dP/d\Xi$ depends only on $\Delta\omega$ and $\Delta\underline{k}$ breaks down far from resonance, one may nevertheless anticipate some of the effects arising from resonance conditions for the case of scattering by polaritons. These would be expected to occur principally from the modification of $\Delta\underline{k}$ from the bare phonon momentum transfer to that of polaritons. Since $\Delta\underline{k}$ may then become very large near to resonance, the scattering may transform from polar type to nonpolar as ω or ω' is tuned through resonance. Of course, when the attenuation of the modes $\underline{k}$ and $\underline{k}'$ are significant, then the spread in wave vectors induced by attenuation must also be accounted for, as discussed previously in Section 3.3. Detailed experiments and calculations of this type for resonant scattering by polaritons have yet to be performed.

4. Concluding Remarks

In the preceding we have reviewed two polariton approaches to the theory of the first order RRS in crystals. For most common cases of interest, polariton theory is found to predict resonant scattering properties which are qualitatively, although not necessarily quantitatively, similar to the perturbation theory results (see Section 3.2 and Table 1). Polariton theory thus serves to confirm and validate certain aspects of perturbation theory predictions. The principal differences occur for strongly $\underline{k}$-dependent couplings and weak exciton damping, for which polariton theory predicts markedly enhanced scattering near resonance. It is doubtful, however, whether purely polariton effects can be reliably distinguished within existing experimental data, although it is certainly possible in principle to perform experiments where this can be done. The optimal experiments to be performed for this purpose would be those utilizing resonance with well-separated (for example, Frenkel-type) excitations which also possess large optical oscillator strengths.

Although a polariton formulation is obviously required in principle to account for interference effects associated with multichannel scattering, it has not yet been established whether any such effects have in fact been manifested in existing spectra; they may be difficult to observe experimentally in any case. The appearance of new scattering channels seems to have been observed in the case of Brillouin scattering from a dispersive exciton, where a multiplet of lines is predicted. Nevertheless, to this point in time multichannel and spatial-dispersion effects in RRS still remain relatively unexplored topics for experimental investigation.

Another area where polariton-mediated effects are involved is scattering by TO phonon polaritons. When ω or ω' are tuned through resonance, the transferred momentum $\Delta\underline{k}$ can become large, thus transforming the nature of the scattering from polar to nonpolar (see Section 3.4). Again, these effects have yet to be investigated in detail on an experimental basis.

Turning briefly to the theoretical side, a variety of significant problems deserve further attention, especially with respect to the origins of the differences between, and difficulties with, App. I and II. For example, the question of the effects of damping on the scattering efficiency, and the elimination of the singular behavior stemming from the polariton velocity for undamped, dispersionless excitons, are matters which merit further attention within App. I. For App. II, the appearance of singularities in the scattering amplitude for the undamped, dispersionless exciton case should be addressed further. One more area worthy of pursuit is the polariton formulation of higher-phonon scattering rates. Here certain economies in complexity may be achieved from a reduction of the order of perturbation theory needed to describe the processes under consideration (from third to first, fourth to second, etc.). And although many of the experimental and theoretical considerations outlined in this paper may appear somewhat subtle in nature, they nevertheless provide a variety of interesting, and certainly challenging, opportunities for further investigation.

Acknowledgements

The author thanks Drs. J.L. Birman, R.K. Chang, A.K. Ganguly, D.L. Mills, and R. Zeyher for enlightening discussions.

References

Agranovich, V.M. (1960): Sov. Phys.-JETP 10, 207
Agranovich, V.M., V.L. Ginsburg (1966): *Spatial Dispersion in Crystal Optics and the Theory of Excitons* (Interscience, New York)
Agrawal, G.S., D.N. Pattanayak, E. Wolf (1971): Phys. Rev. Lett. 27, 1022
Balkanski, M. (ed.) (1971): *Light Scattering in Solids* (Flammarion, Paris)
Balkanski, M., R.C. Leite, S.P. Porto (eds.) (1976): *Light Scattering in Solids* (Flammarion, Paris)
Balkanski, M. J. Reydellet, D. Trivich (1972): Solid State Commun. 10, 1271
Barker, A.S., R. Loudon (1972): Rev. Mod. Phys. 44, 18
Bendow, B. (1970): Phys. Rev. B2, 5051

Bendow, B. (1971a): Phys. Rev. B4, 552
Bendow, B. (1971b): Phys. Rev. B4, 3489
Bendow, B., J.L. Birman (1970): Phys. Rev. B1, 1678
Bendow, B., J.L. Birman (1971a): In *Light Scattering Spectra in Solids*, M. Balkanski, (ed.) (Flammarion, Paris)
Bendow, B., J.L. Birman (1971b): Phys. Rev. B4, 569
Bendow, B., J.L. Birman, V.M. Agranovich (eds.) (1976): *Theory of Light Scattering in Condensed Matter* (Plenum, New York)
Bendow, B., J.L. Birman, A.K. Ganguly, T.C. Damen, R.C. Leite, J.F. Scott (1970): Opt. Commun, 1, 267
Benoit a la Guillaume, A. Bonnot, J.M. Debever (1970): Phys. Rev. Lett. 24, 1235
Benson, H.J., D.L. Mills (1970): Phys. Rev. B1, 4835
Birman, J.L., J.J. Sein (1972): Phys. Rev. B6, 2482
Born, M., K. Huang (1956): *Dynamical Theory of Crystal Lattices* (Oxford UP)
Brenig, W., R. Zeyher, J.L. Birman (1972): Phys. Rev. B6, 4617
Brillouin, L. (1960): *Wave Propagation and Group Velocity* (Academic, New York)
Brodin, M.S., S.I. Pekar (1960): Sov. Phys.-JETP 11, 55; 1286
Burstein, E., F. DeMartini, (eds.) (1974): *Polaritons* (Pergamon Press, New York)
Claus, R., L. Merten, J. Brandmüller (1975): *Light Scattering by Phonon-Polaritons*, in Springer Tracts in Modern Physics, Vol. 75 (Springer, Berlin, Heidelberg, New York)
Compaans, A. (1973): Prive Communication
Compaans, A., H.Z. Cummins (1973): Phys. Rev. Lett. 31, 41
Dervisch, A., R. Loudon (1976): J. Phys. C. 9, L669
Fano, U. (1956): Phys. Rev. 103, 1702
Faust, W.L., C.H. Henry (1966): Phys. Rev. Lett. 17, 1265
Ferrari, C.A., J.B. Salzberg, R. Luzzi (1974): Solid State Commun. 6, 1081
Ganguly, A.K., J.L. Birman (1967): Phys. Rev. 162, 806
Giallorenzi, T.G. (1971): In *Light Scattering Spectra in Solids*, M. Balkanski (ed.) (Flammarion, Paris)
Gorban, I.S., V.B. Tomifeev (1967): Sov. Phys. - Doklady 6, 878
Hopfield, J.J. (1958): Phys. Rev. 112, 1555
Hopfield, J.J. (1969): Phys. Rev. 182, 945
Huang, K. (1951): Nature 167, 779
Huang, K. (1951): Proc. Roy. Soc. (London) A 208, 352
Kiselev, V.A., B.S. Razbirin, I.N. Ural'tsev (1974): In *Proc. 12th Conf. Phys. Semicon.*, M. Pilkuhn, (ed.) (Teubner, Stuttgart)
Loudon, R. (1963): Proc. Roy Soc. (London) A 275, 218
Loudon, R. (1964): Adv. Phys. 13, 423
Loudon, R. (1965): J. Phys. 26, 677
Lax. M., D.F. Nelson (1974): In *Polaritons*, E. Burstein, F. DeMartini (eds.) (Pergamon, New York)
Lax, M., D.F. Nelson (1975): J. Opt. Soc. Amer. 65, 668
Maradudin, A.A., D.L. Mills (1973): Phys. Rev. B7, 2878
Martin, R.M. (1971): Phys. Rev. B4, 3676
Martin, R.M. (1974): Phys. Rev. B10, 2620
Martin, R.M., T.C. Damen (1971): Phys. Rev. Lett. 26, 86
Martin, R.M., L.M. Falicov (1975): In *Light Scattering in Solids*, M. Cardona (ed.), Topics in Applied Physics, Vol. 8 (Springer, Berlin, Heidelberg, New York)
Marroyannis, C. (1967): J. Math. Phys. 8, 1515; 1522
Messiah, A. (1962): *Quantum Mechanics* (North-Holland, Amsterdam) Chap. XIX
Mills, D.L., A.A. Maradudin, E. Burstein (1968): Phys. Rev. Lett. 21, 1178
Mills, D.L., E. Burstein (1969): Phys. Rev. 188, 1765
Mills, D.L., E. Burstein (1974): Rep. Prog. Phys. 37, 817
Oka, Y., T. Kushida, T. Murahashi, T. Koda (1973): Techn. Report ISSP, No. 601 (Univ. Tokyo, Japan)
Ovander, L.N. (1962): Sov. Phys. - Sol. State 3, 1737
Ovander, L.N. (1965): Sov. Phys. - Uspekhi 8, 337
Pekar, S.I. (1958): Sov. Phys.-JETP 6, 785
Platzman, P.M., N. Tzoar (1969): Phys. Rev. 182, 510
Poulet, H. (1955): Ann. Phys. (Paris) 10, 908

Ralston, J.M., R.L. Wadsack, R.K. Chang (1970): Phys. Rev. Lett. 25, 814
Richter, W., R. Zeyher (1976): Festkörperprobleme XVI
Schmidt, R. (1975): Phys. Rev. B11, 746
Sein, J.J. (1969): Ph.D. Thesis, New York University (unpublished)
Sein, J.J. (1970): Phys. Lett. 32A, 141
Skettrup, T., I. Balsev (1970): Phys. Stat. Sol. 40, 93
Tait, W.C., R.L. Weiher (1969): Phys. Rev. 178, 1404
Ulbrich, R.G., C. Weisbuch (1977): Phys. Rev. Lett. 38, 865
Verlan, E.M., L.N. Ovander (1967): Sov. Phys. - Sol. State 8, 1939
Winterling, G., E. Koteles (1977): Sol. State Commun. 23, 95
Wright, G.B. (ed.) (1969): *Light Scattering Spectra in Solids* (Springer, Berlin, Heidelberg, New York)
Yu, P.Y., Y.R. Shen, Y. Petroff, L.M. Falicov (1973): Phys. Rev. Lett. 30, 283
Zeyher, R., J.L. Birman (1974): In *Polaritons*, E. Burstein, F. DeMartini (eds.) (Pergamon, New York)
Zeyher, R., W. Brenig, J.L. Birman (1972): Phys. Rev. B6, 4613
Zeyher, R., C.S. Ting, J.L. Birman (1974): Phys. Rev. B10, 1725
Zubarev, D.N. (1960): Sov. Phys. - Uspekhi 3, 320

Springer Series in
Solid-State Sciences

Volume 1
C.P. Slichter

Principles of Magnetic Resonance

2nd revised and expanded edition.
1978. 112 figures. X, 397 pages
ISBN 3-540-08476-2

Contents: Elements of Resonance. – Basis Theory. – Magnetic Dipolar Broadening of Rigid Lattices. – Magnetic Interactions of Nuclei with Electrons. – Spin-Lattice Relaxation and Motional Narrowing of Resonance Lines. – Spin Temperature in Magnetism and in Magnetic Resonance. – Double Resonance. – Advanced Concepts in Pulsed Magnetic Resonance. – Electric Quadrupole Effects. – Electron Spin Resonance. Summary. – Problems. – Appendices.

Volume 2
O. Madelung

Introduction to Solid-State Theory

Translated from the German by B.C. Taylor
1978. 144 figures. Approx. 530 pages
ISBN 3-540-08516-5

Contents: Fundamentals. – The One-Electron Approximation. – Elementary Excitations. – Electron-Phonon Interaction: Transport Phenomena. – Electron-Electron Interaction by Exchange of Virtual Phonons.: Superconductivity. – Interaction with Photons: Optics. – Phonon-Phonon Interaction: Thermal Properties. – Local Description of Solid-State Properties. – Localized States. – Disorder. – Appendix: The Occupation Number Representation.

Volume 3
Z.G. Pinsker

Dynamical Scattering of X-Rays in Crystals

1978. 124 figures, 12 tables.
XII, 511 pages
ISBN 3-540-08564-5

Contents: Wave Equation and Its Solution for Transparent Infinite Crystal. – Transmission of X-Rays Through a Transparent Crystal Plate. Laue Reflection. – X-Ray Scattering in Absorbing Crystal. Laue Reflection. – Poynting's Vectors and the Propagation of X-Ray Wave Energy. – Dynamical Theory in Incident-Spherical-Wave Approximation. – Bragg Reflection of X-Ray I.
Basic Definitions. Coefficients of Absorption: Diffraction in Finite Crystal. – Bragg Reflection of X-Ray. II. Reflection and Transmission Coefficients and Their Integrated Values. – X-Ray Spectrometers. Used in Dynamical Scattering Investigation. Some Results of Experimental Verification of the Theory. – X-Ray Interferometry. Moire Patterns in X-Ray Diffraetion. – Generalized Dynamical Theory of X-Ray Scattering in Perfect and Deformed Crystals. – Dynamical Scattering in the Case of Three Strong Waves and More. – Appendices.

Volume 4

Inelastic Electron Tunneling Spectroscopy

Proceedings of the International Conference, and Symposium on Electron Tunneling, University of Missouri-Columbia, USA
May 25-27, 1977
Editor: T. Wolfram
1978. 126 figures, 7 tables
VIII, 242 pages
ISBN 3-540-08691-9

Contents: Review of Inelastic Electron Tunneling. – Applications of Inelastic Electron Tunneling. – Theoretical Aspects of Electron Tunneling. – Discussions and Comments. – Molecular Adsorption on Non-Metallic Surfaces. – New Applications of IETS. – Elastic Tunneling.

Springer-Verlag
Berlin
Heidelberg
New York

Photoemission in Solids I

General Principles

Editors: M. Cardona, L. Ley

1978. 91 figures, 17 tables, Approx. 305 pages
(Topics in Applied Physics, Volume 26)
ISBN 3-540-08685-4

Photoelectron spectroscopy also referred to as ESCA, XPS and UPS is utilized to investigate the electronic structure of solids. It has experienced an explosion since the late sixties which has gone hand in hand with the commercial availability of spectrometers, the availability of synchrotrons and storage rings as light sources, and the development of our understanding of the nature of solid surfaces. The field of photoemmission in solids is treated in this and a forthcoming volume with special emphasis on volume as opposed to surface effects. The present book deals with the general principles of photoelectron spectroscopy, and includes a survey of the field of instrumentation while presenting the theoretical background required to interpret experimental spectroscopy.

Each chapter was written by one or several of the foremost experts in the field. The book should appeal to physicists and chemists interested in photoelectron spectroscopy or to those trying to enter the field.

The second volume (Topics in Applied Physics, Volume 27) deals with case studies.

Contents:

Springer-Verlag
Berlin
Heidelberg
New York